T0077226

Cambridge Elements

INTELLIGENT METASURFACE SENSORS

Lianlin Li
Peking University
Hanting Zhao
Peking University
Tie Jun Cui
Southeast University, China

Shaftesbury Road, Cambridge CB2 8EA, United Kingdom

One Liberty Plaza, 20th Floor, New York, NY 10006, USA

477 Williamstown Road, Port Melbourne, VIC 3207, Australia

314–321, 3rd Floor, Plot 3, Splendor Forum, Jasola District Centre, New Delhi – 110025, India

103 Penang Road, #05–06/07, Visioncrest Commercial, Singapore 238467

Cambridge University Press is part of Cambridge University Press & Assessment, a department of the University of Cambridge.

We share the University's mission to contribute to society through the pursuit of education, learning and research at the highest international levels of excellence.

www.cambridge.org
Information on this title: www.cambridge.org/9781009454131

DOI: 10.1017/9781009277242

First published 2023

A catalogue record for this publication is available from the British Library

ISBN 978-1-009-45413-1 Hardback
ISBN 978-1-009-27727-3 Paperback
ISSN 2399-7486 (online)
ISSN 2514-3875 (print)

Intelligent Metasurface Sensors

Elements in Emerging Theories and Technologies in Metamaterials

DOI: 10.1017/9781009277242
First published online: December 2023

Lianlin Li
Peking University

Hanting Zhao
Peking University

Tie Jun Cui
Southeast University, China

Author for correspondence: Lianlin Li, lianlin.li@pku.edu.cn

Abstract: Intelligent electromagnetic (EM) sensing is a powerful contactless examination tool in science, engineering, and the military, enabling us to "see" and "understand" visually invisible targets. Using intelligence, the sensor can organize by itself the task-oriented sensing pipeline (data acquisition plus processing) without human intervention. Intelligent metasurface sensors, synergizing ultrathin artificial materials (AMs) for flexible wave manipulation and artificial intelligences (AIs) for powerful data manipulation, emerge in response to the proper time and conditions and have attracted growing interest over the past several years. The authors expect that the results in this Element could be utilized to achieve goals that conventional sensors cannot achieve, and that the developed strategies can be extended over the entire EM spectrum and beyond, which will have a major impact on the society of the robot–human alliance.

Keywords: electromagnetic sensing, wireless sensing, artificial intelligence, artificial materials, metamaterials and metasurfaces

ISBNs: 9781009454131 (HB), 9781009277273 (PB), 9781009277242 (OC)
ISSNs: 2399-7486 (online), 2514-3875 (print)

Contents

1 Introduction

Intelligent electromagnetic (EM) sensing is a powerful, nondestructive examination tool in science, engineering, and the military, enabling us to "see" and "understand" visually invisible targets. By "intelligent" here we mean that the sensor can organize by itself the task-oriented sensing pipeline (data acquisition plus processing) without human intervention. Although some EM sensing schemes, for instance, real aperture, synthetic aperture, and coding aperture, have been developed, they struggle in trading off some critical factors affecting the cost–performance index due to lack of intelligence. For instance, the coding-aperture and synthetic-aperture sensing strategies can produce high-quality images in a hardware-cost-efficient manner, but at the cost of computationally inefficient digital data processing. In contrast, the real-aperture sensing strategy can be optimized to be nearly free of data processing but requires a large number of expensive sensors for dummy data acquisition. To tackle these formidable challenges, intelligent metasurface sensors, which synergize ultrathin artificial materials (AMs) for flexible wave manipulation and artificial intelligences (AIs) for powerful data manipulation, are emerging in response to the proper time and conditions, and have attracted growing interest over recent years.

When I (Li) was a researcher in 2006–2008 in the Institute of Electronics, Chinese Academy of Sciences, China, I was particularly interested in beating the diffraction limit by means of a synergizing exploration of physically engineered structures and digital computation algorithms (specifically, sparse-aware reconstruction algorithms). After about four years research in petroleum engineering at Texas A&M University, USA, I, in collaboration with Professor Tie Jun Cui at Southeast University, China, started research in intelligent EM sensing by synergizing metamaterials with machine learning techniques at Peking University. During the past several years, we have made many fruitful efforts and have primarily established three frameworks of intelligent EM sensing: the nearly digital-computing-free EM imager based on the linear-machine-learning-driven reprogrammable metasurface, the hybrid-computing intelligent EM camera based on the deep-learning-driven reprogrammable metasurfaces, and the hybrid-computing intelligent integrated EM sensing system. This is a newly emerging research direction, involving different disciplines of information and sciences. We expect that our results could be utilized to achieve goals that conventional sensors cannot achieve, and that the strategies developed can be extended over the entire EM spectrum and beyond, which will have a major impact on the society of the robot–human alliance. This is one of the major impetuses that prompted us to write this short Element, which we hope will provide many helpful guidelines for our readers, especially graduate students and young researchers.

As mentioned earlier, EM sensing has been widely demonstrated to be an effective yet nondestructive means of gathering data on objects under all-weather, all-time operational conditions, and it serves as a fundamental asset in science, engineering, and the military. Typically, an entire sensing chain consists of two major building parts: data acquisition and data processing. To date, three kinds of EM sensing schemes have been proposed and widely explored in practice: real aperture, synthetic aperture, and coding aperture. Conventionally, EM sensors have to struggle with trading off the cost–performance indexes of data acquisition and data processing, especially when dealing with high-dimensional data or a "data crisis." For instance, the coding-aperture and synthetic-aperture sensing strategy could produce high-quality images with one or a few sensors, but at the cost of computationally inefficient digital computation algorithms. In contrast, the real-aperture strategy can be optimized to have nearly negligible pressure on the digital data processing, but requires costly sensors for the data acquisition on the physical level. This situation becomes more and more serious with the ever-increasing demand for the high-frame-rate three-dimensional (3D) imaging, since it is inevitably accompanied by dramatically increasing data rates that pose a heavy burden on data acquisition, system communication, and subsequent reconstruction algorithms. Thus, it is urgently necessary to establish the framework of intelligent sensing with the relevant information-data acquisition and processing. And so, intelligent metasurface sensors, integrating reprogrammable metamaterials for adaptive wave manipulation on the physical level with machine learning techniques for powerful data processing on the digital level as a unique entity, are emerging and are drawing more attention in the research.

The reprogrammable metasurface, an active research branch of metamaterials and metasurfaces, is designed to be capable of manipulating wave–matter interactions dynamically. By "reprogrammable" we here mean that the metasurface is capable of realizing multiple distinct EM functionalities by manipulating the control digital sequences through an embedded field-programmable gate array (FPGA) instead of by hardware modifications. The reprogrammable metasurface is composed of a two-dimensional (2D) array of elaborately designed elements (called meta-atoms), and the meta-atom is controllable or reprogrammable because it is integrated with tunable materials or active elements. Thereby, the reprogrammable metasurface, when it is properly integrated with a "brain" of advanced algorithms, can be optimized to be intelligent in manipulating wave–matter–information interactions. Thus, it can be envisioned that such an intelligent metasurface could be utilized to radically advance sensing architectures and empower EM sensors.

The body of this Element is arranged as follows:

Section 2 is mainly about the fundamental principles and design of reprogrammable metasurfaces.

Section 3 deals with the basic information theory of reprogrammable metasurfaces.

Section 4 discusses compressive metasurface imagers, including single-frequency multiple-shot compressive imagers and wide-band single-shot compressive imagers.

Section 5 is about machine-learning metasurface imagers.

Section 6 presents the deep-learning-driven metasurface camera. We show that such a camera is capable of monitoring the notable or nonnotable movements of multiple noncooperative persons in a real-world setting, even when it is excited by the stray wireless signals already existing everywhere.

Section 7 deals with intelligent integrated sensing systems. We discuss two recent advancements: variational auto-encoder-based intelligent integrated sensing, and free-energy-based intelligent integrated sensing. We show that all these systems can yield a superior performance compared to conventional sensing strategies that optimize measurement and processing separately.

To summarize, we discuss the emerging direction of intelligent metasurface sensors in this Element, which synergizes ultrathin AMs for flexible wave manipulation on the physical level and AI for powerful data manipulation on the digital level. We expect that the sensors discussed in this Element can be extended over the entire EM spectrum and beyond, a feature that could be utilized to achieve the goal that conventional sensors cannot achieve. This new research direction involves multiple disciplines of information and sciences, and many issues need to be investigated further. We here highlight three potential routes for the further development of intelligent metasurface sensors. First, the most fundamental but still unresolved problem is to address why and when the integration of AMs and AIs works. Second, there is a demand to design more specialized reprogrammable metasurfaces with miniaturization, lower hardware cost, and lower power consumption. Third, it is appealing to extend hybrid-computing-based intelligent sensing systems to all-wave sensing systems in order to overcome the barrier of Moore's law. This would enable users to enjoy many fantastic "green" properties in terms of cost, power consumption, complexity, and efficiency.

2 Reprogrammable Coding Metasurfaces

Reprogrammable coding metasurfaces are evolved from active metamaterials, in which the constituting meta-atoms are represented in the form of the digital information rather than the equivalent medium parameters adopted by conventional metamaterials. The digital representation of metamaterials provides us with digital information insights into metamaterials, bridging the physical world and the digital world. Here, we start our discussion of the historical background and underlying principles of reprogrammable coding metasurfaces. Afterwards, we discuss three recent representative examples of reprogrammable coding metasurfaces: the dynamic hologram, the reprogrammable Orbital Angular Momentum (OAM)-beam generator, and the intelligent wireless energy reallocator of stray Wi-Fi signals. We expect that the reprogrammable coding metasurface holds much promising potential in enabling the design of future smart EM platforms in a variety of applications such as display, security, data storage, and information processing.

2.1 Historical Background and Fundamental Principles

The reprogrammable coding metasurface, a kind of active metamaterial, was developed out of digital coding metasurfaces suggested first by Cui et al. in 2014 [1]. Compared with conventional metamaterials, coding metamaterials exhibit a unique property [1–3]: the building meta-atoms are represented with the digital information rather than the equivalent medium parameters used by conventional metamaterials, as conceptually shown in **Fig. 2.1a**. For instance, a one-bit coding metamaterial is composed of meta-atoms with two distinct EM statuses. In this case, the meta-atom can be assigned binary digital information states: "0" and "1." When the meta-atom is controllable [1–14], for instance, by integrating tunable materials (like VO_2 [e.g., 8], $Ge_2Sb_2Te_5$ [e.g., 11]) or active elements (like PIN diodes [e.g., 1], or by Schottky diodes [e.g., 12]), the resultant coding metasurface is referred to as the reprogrammable coding metasurface. **Figure 2.1b** displays an example of a one-bit reprogrammable meta-atom embedded with a PIN diode [1], which behaves as digital states "0" and "1," corresponding to the EM reflection phases of 0° and 180° respectively, when the PIN diode is switched "OFF" and "ON" respectively. The EM reflection phases of the one-bit meta-atom under the two states are plotted in **Fig. 2.1b** as well, from which we can see that the phase difference reaches 180° at the designed frequency of around 8.6 GHz. The digital representation of metamaterials could provide us with digital information insights into metamaterials, and bridge the physical world and the digital world.

Here, we consider the first presentation of the reprogrammable coding metamaterial [1] to illustrate its basic operational mechanism. The reprogrammable coding metasurface is electronically manipulated to achieve various EM functionalities by configuring the different sequences of binary coding through an FPGA.

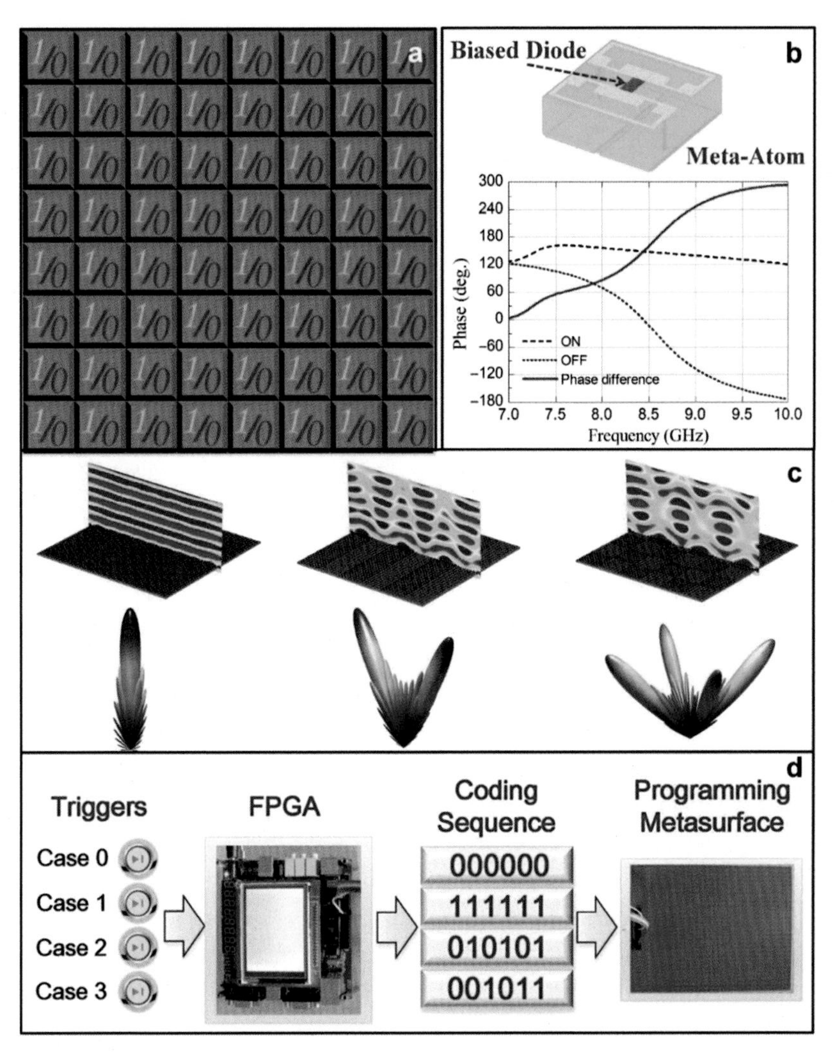

Figure 2.1 The digital coding metamaterials and their constitutive digital meta-atoms. **(a)** The one-bit digital coding metamaterial composed of two types of meta-atoms with "0" and "1" states. **(b)** Top: the structure of a one-bit digital meta-atom, which behaves as the digital states "0" and "1" when the biased PIN diode is switched "OFF" and "ON" respectively. Bottom: the corresponding phase responses of the one-bit digital meta-atom as the PIN diode is switched "OFF" and "ON" over a range of frequencies. **(c)** The theoretical and full-wave simulation results of the one-bit digital coding metamaterials with different coding sequences under the normal incidence of EM waves. **(d)** The configuration and its working mechanism of the programmable metamaterial.

For instance, under the uniform coding sequence of "000000.../000000...," the normally incident plane wave is mainly reflected to a single main beam at broadside, as shown in the left column of **Fig. 2.1c**; under the periodic coding sequence of "010101.../010101...," the normally incident plane wave is mainly reflected to two symmetrical beams, as shown in the middle column of **Fig. 2.1c**; while in the case of the periodic coding sequence of "010101.../101010.../010101.../101010...," the normally incident plane wave is mainly reflected to four symmetrical beams, as shown in the right column of **Fig. 2.1c**. To achieve the real-time reprogrammable control, the different control binary sequences are precalculated and stored in the FPGA, and the FPGA is equipped on the physical level of metamaterial to flexibly trigger the desired coding patterns. As displayed in **Fig. 2.1d**, when the FPGA detects the trigger signal, the relevant bias voltage will be implemented on each diode according to the coding sequences (000000, 111111, 010101, and 001011). Besides these four example coding sequences, we remark that many more coding patterns and functions can be designed using the coding metamaterial theory.

The concept of the reprogrammable coding metamaterial can be extended from one-bit coding to multibit coding. For instance, the structure of a two-bit digital meta-atom is integrated with three PIN diodes [15–16] and has four different phase responses to define the two-bit digital states "00," "01," "10," and "11" with the phase difference of 90°. Similarly, higher-bit digital meta-atoms can be defined and realized. In contrast to the traditional metamaterials, which are controlled by the effective medium parameters, the reprogrammable coding metasurfaces, due to the flexibility and simplicity of coding metamaterials, have been extensively studied in the past years for the achieving of many fruitful applications, for instance, beam manipulation [17], vortex beams [18–20], polarization control [17, 21], holograms [22], computational imaging [23–24], and antenna designs [25].

2.2 Wireless Energy Reallocation of Stray Wi-Fi Signals

We discuss the wireless energy reallocation of ambient stray wireless signals using a reflection-type reprogrammable coding metasurface, proposed by Shuang et al. in 2021 [26]. Taking an illustrative example, we consider the reallocation of the energy of commodity 2.4 GHz Wi-Fi signals with the 802.11n protocol. To this end, Shuang et al. designed a one-bit reprogrammable coding metasurface working at around 2.4 GHz, and developed an efficient optimization algorithm for obtaining the control binary coding sequences of the reprogrammable coding metasurface. Here, selected results are provided to show the performance of the proposed strategy in dynamically reallocating the spatial energy of wireless signals without consuming extra energy. Notice that such a strategy is different from conventional device-oriented or retransmission-oriented techniques of wireless energy reallocation, since they require the expenditure of extra energy due to the use of amplifiers and other associated active devices. It can be expected that the strategy of Shuang et al. could provide us with a promising solution for future smart homes, wireless communications, and so on.

2.2.1 Design of a One-Bit Reprogrammable Coding Metasurface

The one-bit reprogrammable coding metasurface used here is composed of 3×3 metasurface panels, each metasurface panel consists of 8×8 electronically controllable meta-atoms, and each meta-atom is embedded with a PIN diode (SMP1345-079LF). Thus, the whole reprogrammable metasurface has 24×24 independently controllable meta-atoms, and the full size of the metasurface is 1.296 m $\times$ 1.296 m. The metasurface is fabricated using the standard printed circuit board (PCB) technology, and the thickness of the fabricated metasurface is about 3 mm in total, where the direct current (DC) biasing circuit has been included. **Figure 2.2a and b** report the front and back views of the designed one-bit reprogrammable metasurface, respectively. The zoomed version of the meta-atom is inserted in **Fig. 2.2a**, and the FPGA-based micro control unit (MCU) with size of 90 mm $\times$ 90 mm is inserted in **Fig. 2.2b** as well. To achieve the flexible control of PIN diodes of reprogrammable coding metasurface panels, the MCU with the 50 MHz CLK is connected with three metasurface panels through three 1.0 m-length winding wires in parallel, and each of them is connected with two metasurface panels in series, as sketched in **Fig. 2.2c**. A FPGA chip is used to distribute all commands to all PIN diodes. Each metasurface panel is equipped with eight 8-bit shift registers (SN74LV595APW), and eight PIN diodes will be sequentially controlled. As a consequence, the MCU sends the commands over 24 independent branch channels, leading to the real-time manipulation of 576 PIN diodes. Moreover, 576 red-color LEDs are employed to indicate the status of 576 PIN diodes. Here we would like to say that such a control strategy can be readily extended for more PIN diodes by concatenating more metasurface panels, allowing adjustable rearrangement of metasurface panels to meet various needs.

The sketch map of the one-bit electronically controllable meta-atom is illustrated in **Fig. 2.3a and b**. The meta-atom has a sandwich structure: a square patch and a metal ground layer spaced by the F_4BM substrate with relative dielectric constant of 2.55 and loss tangent of 0.0015. Other parameters of the meta-atom are set as follows: the size of the top square patch is

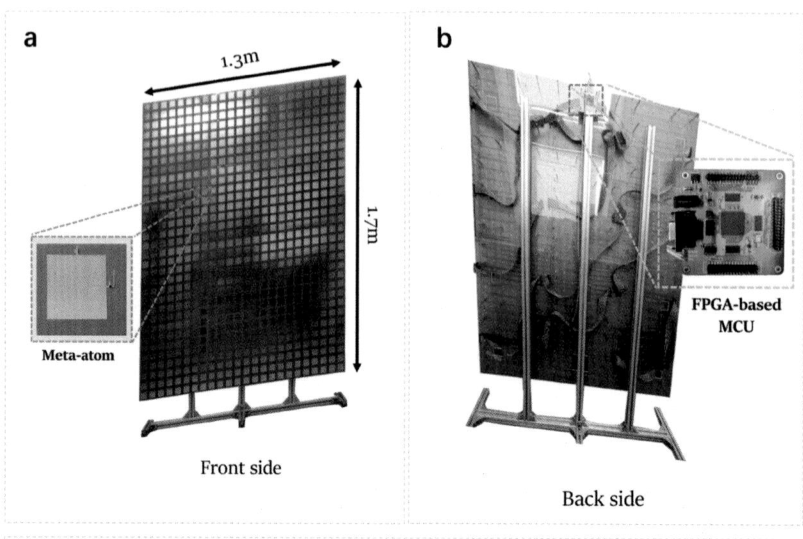

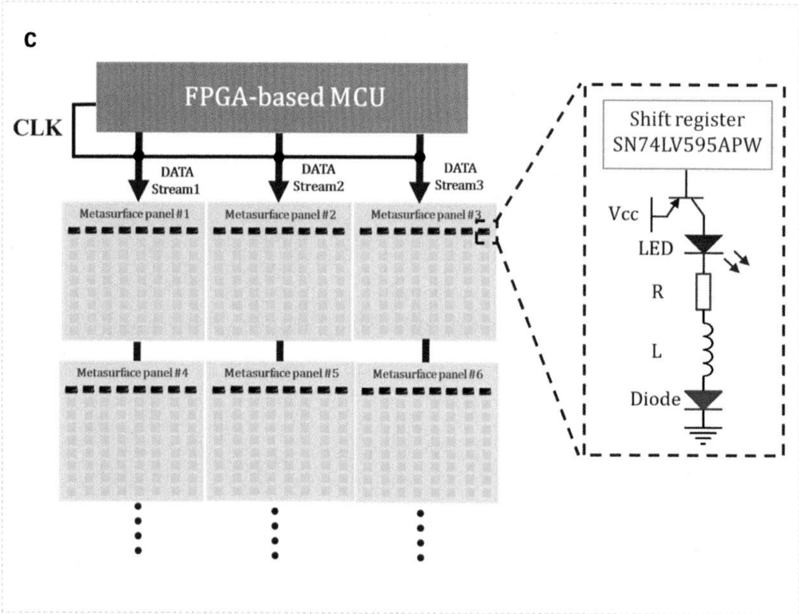

Figure 2.2 The prototype of a fabricated one-bit reprogrammable coding metasurface [23, 26]. **(a)** The front of the metasurface. **(b)** The back of the metasurface. **(c)** The control architecture of the FPGA board and the magnified view of a logic circuit on the metasurface panel.

37 mm × 37 mm, the total length of the 90°-bending strip is 14.6 mm, and the thickness of the substrate F_4BM is 1.5 mm. In order to apply appropriate external DC voltage and associated feeding lines for the PIN diode, another

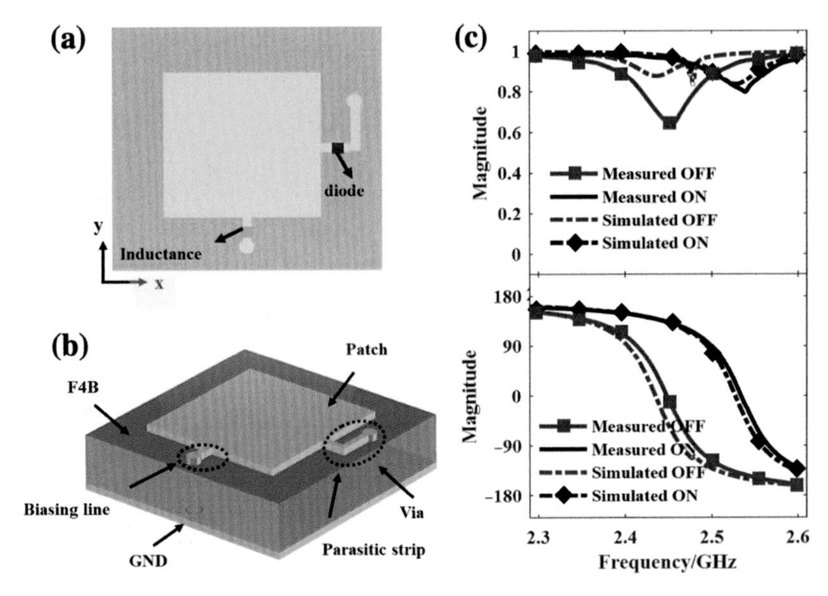

Figure 2.3 The geometrical map and its operational performance of the proposed meta-atom [26]. **(a)** Top view of the meta-atom. **(b)** Perspective view of the meta-atom. **(c)** The simulated and experimental amplitude-frequency response of the designed meta-atom, and the simulated and experimental phase-frequency response of the designed meta-atom.

layer of substrate (FR-4) with thickness of 0.7 mm has been mounted below the ground plane. To achieve the good isolation between the radio-frequency (RF) and DC signals, an inductor (TDK MLK1005S33NJT000) with inductance of L = 33 nH and self-resonance frequency higher than 3.5 GHz has been introduced. Moreover, a current-limiting resistance has also been utilized to guarantee as low a power consumption as possible. A PIN diode with low insertion loss ($\leq$ 0.2 dB) and high isolation ($\geq$ 13 dB) in the 2.4 GHz Wi-Fi frequency band is embedded at the gap between the square patch and the 90°-bending strip. When the applied PIN diode is switched from "ON" (or "OFF") to "OFF" ("ON"), the meta-atom will experience a reflection phase difference of 180° when it is illuminated by a plane wave, as shown in **Fig. 2.3c** for the simulated and experimental results of phase-frequency and amplitude-frequency responses. It can be observed that regardless of the operational states of PIN diode, the simulated reflection amplitudes are larger than 0.8, and the frequency bandwidth within the phase difference 180°±20° is about 55 MHz (2.401–2.456 GHz).

2.2.2 Modified Gerchberg–Saxton (GS) Algorithm

One critical issue of reprogrammable coding metasurfaces is to find the optimal control coding sequence for achieving the desired EM functionality. In other words, it is critical to optimize the binary sequence for controlling the status of PIN diodes, such that the desired EM response of the reprogrammable coding metasurface to the illumination of Wi-Fi signals can be obtained. It is a typical NP-hard combinatorial optimization problem. To address this challenging problem, we made a two-aspect contribution as follows: (1) a more realistic EM model of a meta-atom has been introduced in the context of the effective induced current inversion; (2) the standard Gerchberg–Saxton (GS) algorithm originally developed for treating the continuous optimization problem [27] has been modified with more careful back/forward propagation operations. More details are provided in the following discussion.

EM Modeling of Meta-atom and Metasurface In light of the well-known Huygens' principle, the EM response of the meta-atom can be accurately obtained once its induced equivalent current is obtained. Taking this fact into account, Shuang et al. proposed to divide uniformly the meta-atom into P × Q grids, and the induced current within each grid with an area of Δ was approximated to be uniform. The relationship between the scattering field **E** of each meta-atom and the induced equivalent currents **J** can be represented as

$$\mathbf{E}(\mathbf{r}) = \sum_{\mathrm{px}=1}^{\mathrm{P}} \sum_{\mathrm{py}=1}^{\mathrm{Q}} \mathrm{i}\omega\mu \int_{\Delta} \overline{\mathbf{G}}\left(\mathbf{r}, \mathbf{r}'_{\mathrm{px,py}} + \delta\mathbf{r}'\right) \cdot \mathbf{J}\left(\mathbf{r}'_{\mathrm{px,py}} + \delta\mathbf{r}'\right) \, \mathrm{d}\delta\mathbf{r}'. \tag{2.1}$$

Herein, $\overline{\mathbf{G}}\left(\mathbf{r}, \mathbf{r}'\right) = \left[\bar{\mathrm{I}} + \frac{\nabla\nabla}{k_0^2}\right] \frac{e^{\mathrm{j}k_0|\mathbf{r}-\mathbf{r}'|}}{4\pi|\mathbf{r}-\mathbf{r}'|}$ denotes the Dyadic Green's function in free space. Moreover, $\mathbf{r}'_{\mathrm{px,py}}$ denotes the central coordinate of the $(\mathrm{px}, \mathrm{py})$-th grid, and k_0 is wavenumber. The integration in **Eq. (2.1)** is implemented over the grid. When the grid's area tends to 0, namely $\Delta \to 0$, the scattering field **E** becomes

$$\mathrm{i}\omega\mu\Delta \sum_{\mathrm{px}=1}^{\mathrm{P}} \sum_{\mathrm{py}=1}^{\mathrm{Q}} \overline{\mathrm{G}}\left(\mathbf{r}, \mathbf{r}'_{\mathrm{px,py}}\right) \cdot \mathbf{J}\left(\mathbf{r}'_{\mathrm{px,py}}\right).$$

For numerical implementations, **Eq. (2.1)** can be reformulated in a compact form, namely

$$\mathbf{E} = \begin{bmatrix} \mathrm{E}^{\mathrm{x}}_{\mathbf{r}} \\ \mathrm{E}^{\mathrm{y}}_{\mathbf{r}} \end{bmatrix} = \begin{bmatrix} \mathrm{A}^{\mathrm{xx}}_{\mathbf{r},\mathbf{r}'} & \mathrm{A}^{\mathrm{xy}}_{\mathbf{r},\mathbf{r}'} \\ \mathrm{A}^{\mathrm{yx}}_{\mathbf{r},\mathbf{r}'} & \mathrm{A}^{\mathrm{yy}}_{\mathbf{r},\mathbf{r}'} \end{bmatrix} \begin{bmatrix} \mathrm{J}^{\mathrm{x}}_{\mathbf{r}'} \\ \mathrm{J}^{\mathrm{y}}_{\mathbf{r}'} \end{bmatrix} = \mathbf{AJ}. \tag{2.2}$$

Here, $\mathbf{r}'$ goes over the central coordinate of all grids, $\mathbf{r}$ denotes the observation position, $\mathbf{A}$ is a mapping matrix with entries coming from the Dyadic Green's function, and the induced current is organized into a $2 \times P \times Q$ column vector $\mathbf{J}$. Then, a model of that element at either the ON or the OFF state illuminated by a plane wave is calculated through numerical simulation, and the resultant copolarized scattering field is collected at a distance far enough from the element, arranged into a column vector $\mathbf{E}$. Now, the least-square method can be used to retrieve the induced equivalent current $\mathbf{J}$, namely

$$\mathbf{J} = \left(\mathbf{A}'\mathbf{A} + \gamma\mathbf{I}\right)^{+}\mathbf{A}'\mathbf{E}, \tag{2.3}$$

where γ denotes an artificial regularization parameter, $\mathbf{I}$ is the unit matrix, and $+$ presents the matrix pseudo inverse. In our implementations, the surface of the meta-atom was uniformly divided into 10×10 square grids with area of $0.044\lambda \times 0.044\lambda$, and the scattering field $\mathbf{E}$ was numerically obtained at the distance of 0.5λ away from the meta-atom illuminated by the plane wave with the intensity of 1 V/m. Note that, unless otherwise stated, the full-wave simulations are conducted by implementing a commercial software, CST, throughout this Element. The equivalent induced current of a meta-atom at 2.45 GHz can be calculated according to **Eq. (2.3)**, where $\gamma = 10^{10}$ is used.

Then, we can synthesize the EM field induced from the whole reprogrammable coding metasurface using the superposition principle. For simplicity, the electrical field from a Wi-Fi router is considered, which can be approximated with a spherical wave, namely

$$E^{in} = |E^{in}|\exp(j\varphi_{in}) = A_0 \cos^{q}(\theta_f)\frac{\exp(jk_0 r_f)}{r_f}, \tag{2.4}$$

where A_0 is a calibration constant, and θ_f and r_f denote the elevation angle and the observation distance in the spherical coordinate system centered at the Wi-Fi router, respectively. Here, the EM interaction among different meta-atoms has been ignored due to the introduction of macro meta-atoms. As a consequence, the electrical field scattered from the whole metasurface aperture illuminated by a Wi-Fi router can be expressed as following, namely

$$E_s(r) = \sum_{n_x=1}^{N}\sum_{n_y=1}^{M} A(\mathbf{r}, \mathbf{r}_{n_x,n_y})E^{in}_{n_x,n_y} \odot J^{ON/OFF}_{n_x,n_y}. \tag{2.5}$$

Here, the summation in **Eq. (2.5)** is performed over all meta-atoms of the reprogrammable coding metasurface, where (n_x, n_y) denotes the running indices of all elements along the x and y directions, $E^{in}_{n_x,n_y}$ and $\mathrm{J}^{\mathrm{ON/OFF}}_{\mathrm{n_y,n_y}}$ denote the incident Wi-Fi wave and the induced current on the (n_x, n_y)-th element, respectively, and $\odot$ denotes the element-wise product. The flowchart of the proposed algorithm procedure has been provided in **Fig. 2.4a**. Here, we would like to say that under the same computation configuration, the method just outlined will cost only a few minutes; however, it will take tens of hours using the CST microwave studio package in a typical personal computer.

Other Details of the Modified GS Algorithm The GS algorithm, with some modifications, can be applied to find the optimal coding sequence of a one-bit reprogrammable coding metasurface such that the desired distribution of spatial energy of ambient Wi-Fi signals can be achieved. Mathematically, the modified GS algorithm is performed to minimize the following objective function:

$$obj = \sum_{i}^{p_x} \sum_{j}^{p_y} \left(E_{goal}(i,j) - \left|E_s(i,j) + E_{in}(i,j)\right|\right)^2. \tag{2.6}$$

Here, $E_{goal}(i,j)$ represents the desired spatial intensity distribution in an observation plane. The complete flowchart of GS algorithm procedure has been summarized in **Fig. 2.4b**. It is apparent that such an optimization strategy behaves in an iterative way, where a binary quantization operation is needed each iteration, as detailed in **Fig. 2.4b**. As for its initial guess, the meta-atoms of programmable coding metasurface are randomly set to be ON or OFF states. After a few tens of iterations in several seconds, the stable convergence can be achieved. Usually, the solution by modified GS algorithm traps into a local minimum. It is worth mentioning that a global optimization algorithm such as the particle swarm optimization (PSO) algorithm or the genetic algorithm (GA) can be further employed to improve the output result by starting with the initial arrangement obtained by using the modified GS algorithm.

2.2.3 Results and Discussions

Here, selected results from [26] are provided to show the performance of the method discussed above. First, the Wi-Fi signal was mimicked by using a vector network analyzer (VNA, Agilent E5071C), and the so-called near-field scanning technique was applied to obtain the spatial distribution of the Wi-Fi signal. More specifically, the VNA was connected with a Wi-Fi transmitting antenna at the location of B (0, –0.291 m, 0.8 m) and an open-ended waveguide scanning probe to scan the spatial distribution of Wi-Fi signal by measuring the transmission coefficient S_{12} at an observation plane. The scanning area is a 0.945 m × 0.945 m square with a sampling space of 0.03 m. We consider six different control coding sequences of reprogrammable coding metasurface, as shown in the top row of **Fig. 2.5**. Accordingly, the simulated spatial

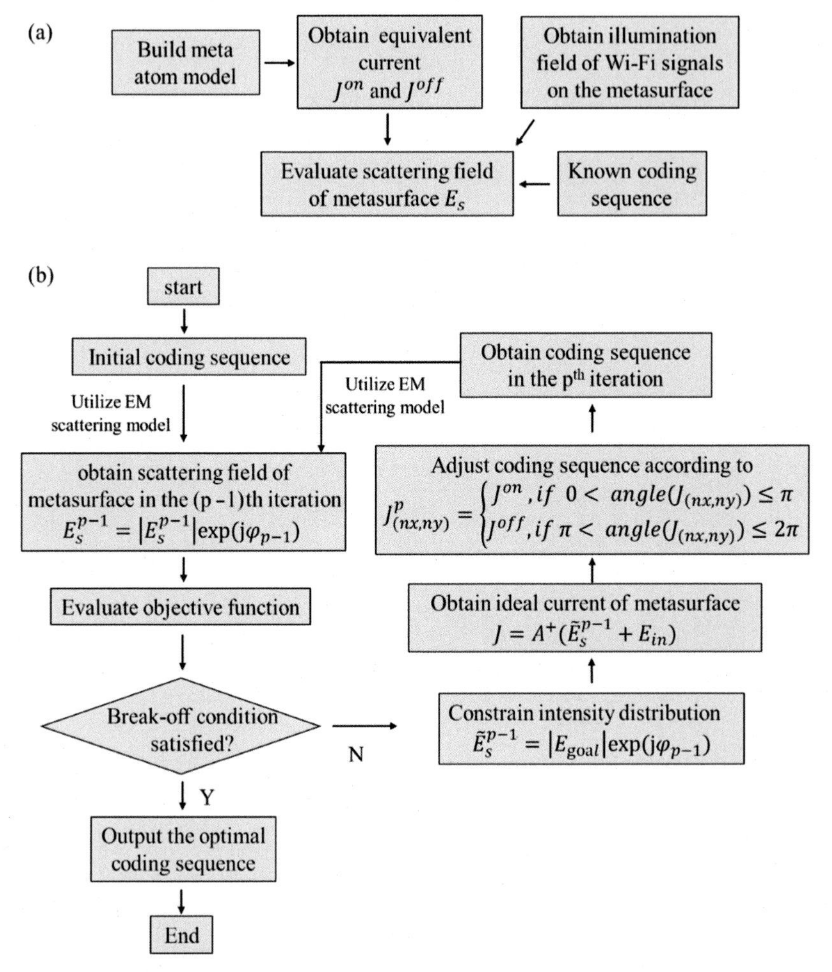

Figure 2.4 Flowchart of the modified GS algorithm for finding the control digital coding pattern of the one-bit reprogrammable coding metasurface [26]. **(a)** EM-scattering field model of the binary-phase electronically controllable metasurface illuminated by Wi-Fi signals. **(b)** Modified GS algorithm for finding the optimal coding sequence of the programmable coding metasurface

distributions of mimicked Wi-Fi signals at three observation planes of $z = 0.984$ m, $z = 1.257$ m, and $z = 1.531$ m away from the one-bit reprogrammable coding metasurface are plotted in the second to fourth rows of **Fig. 2.5**, respectively. For comparison, the corresponding experimental results at the observation plane of $z = 1.531$ m away from the metasurface are also provided in the bottom row of **Fig. 2.5**. The experimental results at the plane of $z = 1.531$ m are obtained, which show good agreement with the theoretical results at the identical plane and validate the effectiveness of the proposed

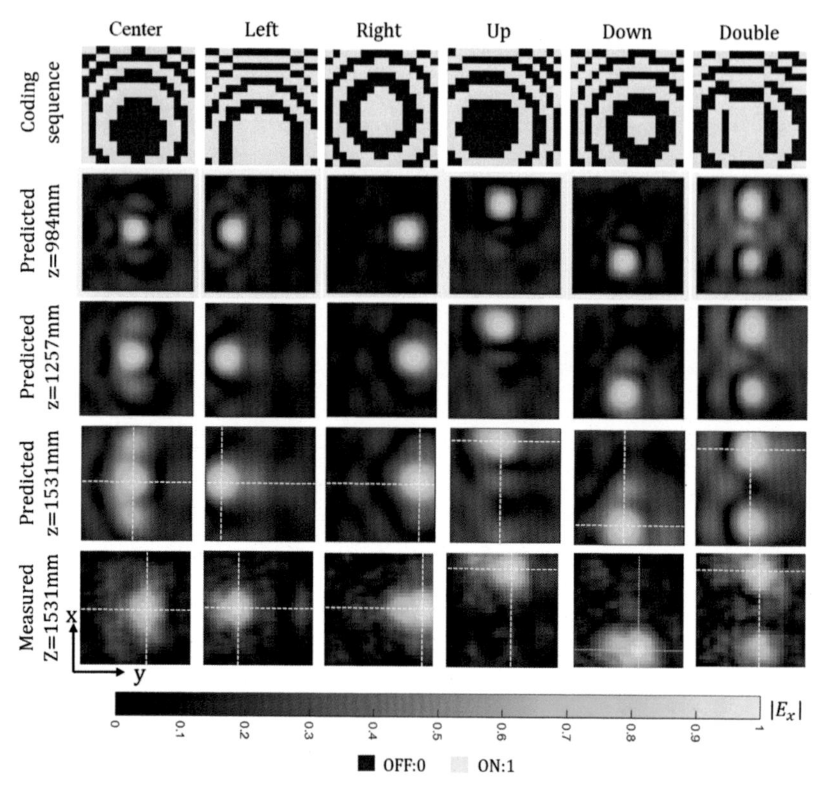

Figure 2.5 Results comparison between prediction and measurement [26]. The first row shows the coding sequences. The second to fourth rows are the optimized normalized spatial intensity distributions predicted respectively at $z = 0.984$ m, $z = 1.257$ m, and $z = 1.531$ m using the proposed method. The fifth row presents the measured normalized spatial intensity distributions at $z = 1.531$ m using the near-field scanning technology.

method. Note that, in these figures, the electrical field distributions have been normalized by their own maximums. Clearly, the predicted energy allocations for all given coding sequences are gradually becoming distorted with the growth of observation plane distance, which is sensible since the coding sequences have been particularly optimized at the given plane $z = 0.984$ m for the aforementioned desired field distributions. Of course, the improved performance of energy allocation could be achieved if adaptive optimizations are performed according to the position of observation distance.

Second, several experimental results are selected from [26] to demonstrate the feasibility and performance of the proposed method for reallocating the spatial energy of commodity Wi-Fi signals. The experiment setup has four

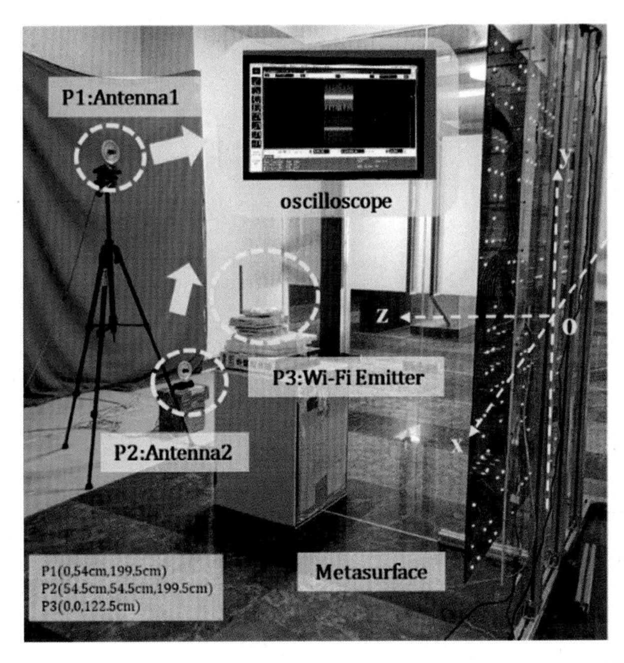

Figure 2.6 Experimental setup of the metasurface-based energy reallocation of commodity Wi-Fi signals [26].

building components: a reprogrammable coding metasurface, a commercial 802.11 n Wi-Fi router, a pair of parabolic antennas, and an oscilloscope (Agilent™ MSO9404A), as shown in **Fig. 2.6**. The Wi-Fi router, working at the seventh operational channel, namely 2.431–2.453 GHz, is randomly placed somewhere in front of the metasurface, say, P3 (0, 0, 1.225 m). Two antennas connected with two ports of the oscilloscope are deployed to acquire the Wi-Fi signals bounced off the reprogrammable coding metasurface. Moreover, the oscilloscope is set with a sampling rate of 10 GHz and a trigger level of 0.02 V.

We examined the performance of the designed programmable coding metasurface and associated optimization algorithm in focusing the commodity Wi-Fi signals at multiple locations, say, P1 (0, 0.54 m, 1.995 m) and P2 (–0.545 m, 0.545 m, 1.995 m), as marked in **Fig. 2.6**. Here, we consider three cases, namely focusing at a single position P1, focusing at a single position P2, and focusing simultaneously at two positions P1 and P2. The control digital coding sequences of the reprogrammable coding metasurface corresponding to these three scenarios are shown in the first row of **Fig. 2.7**. Correspondingly, the spatial distributions of normalized intensity of Wi-Fi signals are plotted in the second row of **Fig. 2.7**, which are numerically obtained by using the numerical method we have developed. The time-domain Wi-Fi signals within the duration of 2 μs,

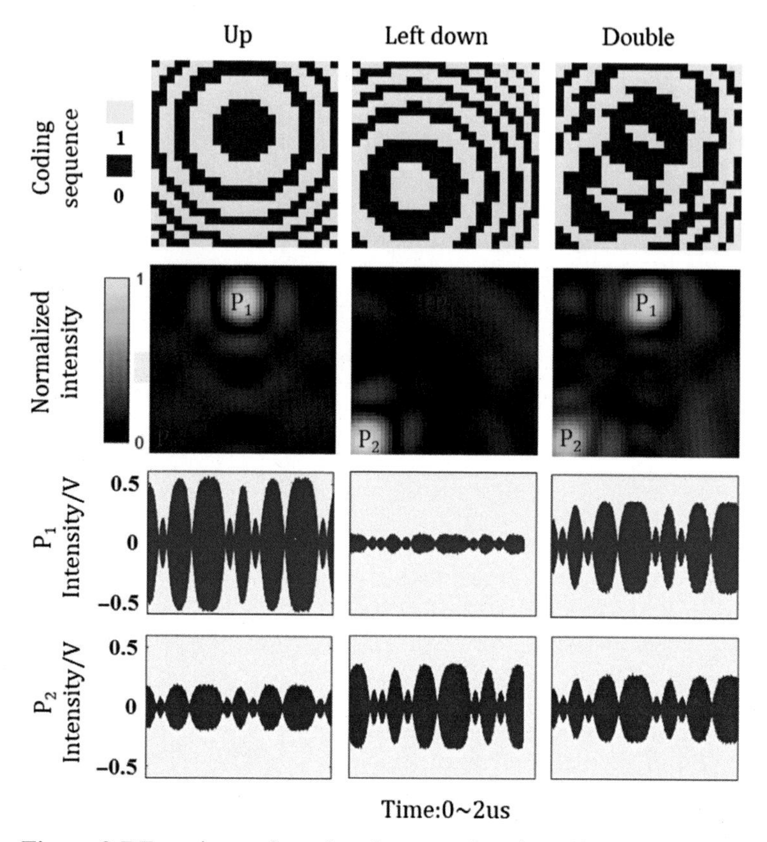

Figure 2.7 Experimental results of metasurface-based wireless energy reallocation of commodity Wi-Fi signals [26]. The first row shows the coding sequences optimized at the z = 1.995 m plane. The second row presents the normalized spatial intensity distributions predicted at the z = 1.995 m plane using the proposed method. The third to fourth rows are the captured commercial Wi-Fi signals at different points P1 and P2.

captured by two receiving antennas at P1 and P2, are depicted in the third and fourth rows of **Fig. 2.7**, respectively. As indicated, the average power of measured Wi-Fi signals at the focused point P1 for the first coding pattern is larger than the one at focused point P2 for the second coding sequence. This makes sense since the location of P2 is farther away from the Wi-Fi router and metasurface than that of P1. In this way, we can conclude that the method of Shuang et al. is capable of supporting the wireless energy reallocation of commodity Wi-Fi signals in a controllable way.

To summarize, we discuss one interesting application of the one-bit reprogrammable coding metasurface in flexibly manipulating the stray wireless signals already existing in our daily lives. We show by experimental results that the developed strategy works very well for controlling the spatial energy of Wi-Fi signals of the IEEE 802.11 n protocol in a flexible and dynamic manner. Such a technique could find some important potential applications in several scenarios. For instance, it can be used to enhance or diminish the received Wi-Fi signals at the intended destination, and resolve the highly concerning problems of the path loss and multipath fading effects in conventional wireless communication systems. Moreover, it can be utilized to build a smart radio environment, which can manipulate the digital information on the physical layer.

2.3 Dynamic Metasurface Holograms

Holography is one of the promising techniques for recording the amplitude and phase information of light to reconstruct the image of objects [28–34]. Over the past decades, a vast number of metasurface holograms have been proposed in the terahertz, infrared, and visible regimes to achieve the holographic images of high efficiency, good image quality, and full color [30–34]. Among them, the reprogrammable metasurface hologram is uniquely positioned due to its unique property: being rewritable. In this section, we take as an illustrative example the reprogrammable metasurface hologram developed by Li et al. [22] to discuss the relevant principle.

2.3.1 Design of Dynamic Metasurface Hologram

The dynamic metasurface hologram by Li et al. relies on a one-bit reprogrammable coding metasurface working in microwave, as conceptually shown in **Fig. 2.8.** The one-bit reprogrammable coding metasurface is composed of 20×20 macro meta-atoms, covering an area of $600 \times 600\ \text{mm}^2$. Here, what we mean by the macro meta-atoms is that a set of adjacent meta-atoms (here, 5×5 meta-atoms) are grouped to form a half-wavelength super cell in order to suppress the corner-related scattering effect. As for the meta-atom, two planar symmetrically patterned metallic structures are printed on the top surface of the F4B substrate with a dielectric constant of 2.65 and a loss tangent of 0.001, with a PIN diode loaded between them, as shown in **Fig. 2.9a**. When the PIN diode is at the state of "ON" (or "OFF") with a biased voltage of 3.3 V (or 0 V), and the corresponding equivalent circuit is illustrated in **Fig. 2.9b**. Then, each meta-atom can be independently programmed to realize the required "0" or "1" state by controlling the applied voltage bias of the diode. **Figure 2.9c and d** show the reflection responses of the meta-atom, which is achieved by using the CST S-parameter simulation. From this set of results, one can see that the theoretical reflection efficiency of the metasurface can reach above 90 percent, and that

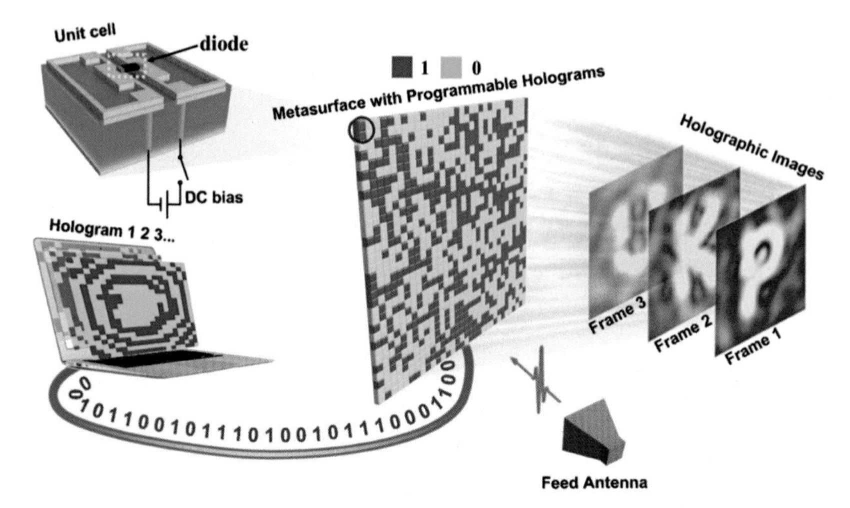

Figure 2.8 Sketch map of a dynamic metasurface hologram. A computer digitally controls the metasurface by dynamically changing the phase distribution (Hologram 1, 2, 3 . . .) computed from the modified GS algorithm. Under the illumination of a feed antenna (on the bottom right side), the metasurface hologram can successively project the holographic images (Frame 1, 2, 3 . . .) at the imaging plane (Z_r), showing the letters "P," "K," and "U."

when the state of the metasurface meta-atom is switched from "1" to "0" (or from "0" to "1"), the reflection amplitudes remain almost unchanged, while the reflection phases are flipped approximately by 180°. By toggling different applied voltages to control the "ON" and "OFF" states of the diodes, the desired distributions of "0" and "1" meta-atoms across the dynamic metasurface hologram can be achieved. As for the control pattern of the one-coding reprogrammable coding metasurface, the modified GS algorithm discussed in the previous section can be applied.

2.3.2 Results

The reconfigurable ability of dynamic metasurface hologram has been experimentally validated at microwave frequencies, where a sequence of holographic images, namely a sentence of "LOVE PKU! SEU! NUS!", were generated. In our experiments, a feeding antenna with working bandwidth from 6 to 14 GHz was employed to generate the x-polarized quasi-plane waves for illuminating the coding metasurface, and a standard waveguide probe was used to scan the image plane with a resolution of 5 × 5 mm^2 to obtain the holographic images.

The phase profiles of the holograms for the letters obtained by the modified GS algorithm are plotted in the left of **Fig. 2.10**, in which two markers with

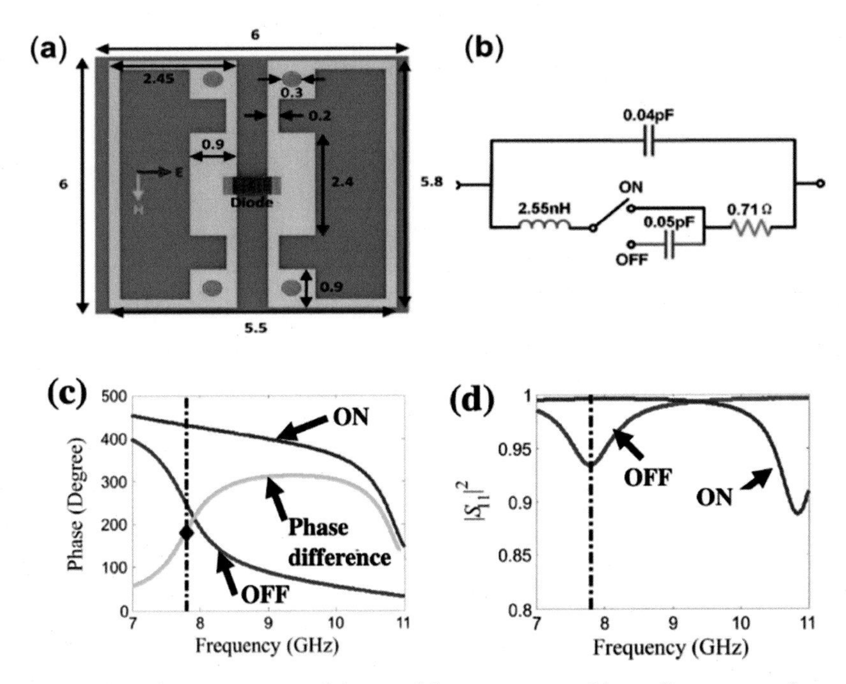

Figure 2.9 The meta-atom of the one-bit reprogrammable coding metasurface and the equivalent circuit models of the diode [22]. **(a)** Top view of the unit cell with detailed geometrical parameters (unit: mm). **(b)** The equivalent circuit models of the diode biased at the "ON" and "OFF" states. **(c)** The phase responses of the metamaterial supercell as the biased diode is "OFF" and "ON" over a range of frequencies. **(d)** The reflection efficiency of the meta-atom when the biased diode is at the states of "OFF" and "ON."

differently shaded values represent two different digital states with constant amplitude. The corresponding experimental results are plotted in the right of **Fig. 2.10**. In experiments, the holograms were illuminated by an x-polarized plane wave, and the holographic image was detected at $Z_r = 400$ mm, which clearly shows the letters "LOVE PKU! SEU! NUS!". An overall efficiency of ~60 percent (defined as the fraction of the incident energy that contributes to the holographic image) was achieved, which was lower than the simulated values (above ~90 percent) due to the phase-quantization loss of the non-ideal plane-wave illumination, the imperfect quality of the commercial PIN-diode, and so on.

2.4 Reprogrammable OAM-Beam Generator

Since the discovery of optical vortex beams in the 1900s, EM waves carrying OAM have been widely investigated [35–43], since their helical phase structures could potentially provide additional degrees of freedom of EM

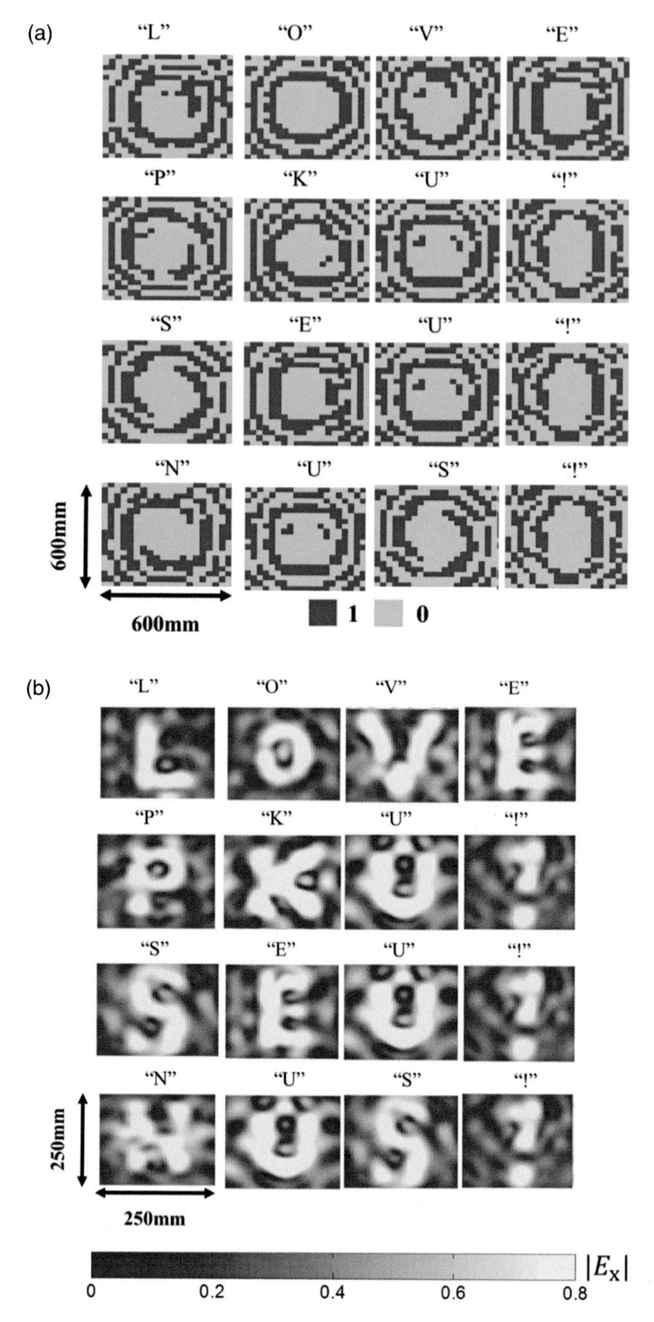

Figure 2.10 **(a)** The binary phase profiles of the coding metasurface for different holographic images [22]. The binary holograms for a sequence of letters of "LOVE PKU! SEU! NUS!" are generated by the modified GS algorithm, which could be realized by switching the states of corresponding unit cells of the coding metasurface. **(b)** The experimental results of holograph images. The experimentally observed holographic images (E_x-field intensity) of "LOVE PKU! SEU! NUS!" measured at the image plane of $Z_r = 0.4$ m.

information, exhibiting potentials in remarkably enhancing the information transfer rate and communication capacity over the existing systems. Roughly, a harmonic OAM beam $E(\rho, z)$ with a topological charge ℓ propagating along the z-direction in free space can be represented as: $E(\rho, z) \propto J_{|\ell|}(k\rho)\exp(j\ell\phi + jkz)$, where $j = \sqrt{-1}$, $J_\ell(\cdot)$ denotes the ℓ-order Bessel function, and k denotes the wavenumber in vacuum. Additionally, ρ and ϕ indicate the radius and angle in the polar coordinate system in the transverse plane perpendicular to the propagation direction, respectively. In this representation, it has been assumed that the vortex center is the same as the coordinate origin. In principle, the OAM-carrying beams with different topological charges are orthogonal to each other, allowing them to be compactly multiplexed together and demultiplexed at low crosstalk.

By now, some reflection-type and transmission-type metasurface devices have been proposed for generating OAM-carrying beams [32–34]. Despite these achievements, they are difficult to be deployed in real-world wireless communication setting. Most of these devices are lack of the reconfigurability, which means that only one specific OAM beam can be generated once the metasurface is fabricated, because the phase and magnitude profiles on the fabricated metasurface are fixed. In order to address this difficulty, several reconfigurable OAM-beam metasurface generators have been proposed recently [e.g., 16, 19, 20]. In this section, we discuss the inexpensive two-bit reprogrammable coding metasurface for generating the OAM-carrying beams with controllable topological charges and vortex centers, which was proposed by Shuang et al. recently. We expect that such a device will offer us a fundamentally new perspective on designing wireless communication architectures at various frequencies, and beyond.

2.4.1 Design of Two-Bit Reprogrammable Coding Metasurface

The two-bit reprogrammable coding metasurface considered here is composed of 48×48 electronically controllable digital meta-atoms, as shown in **Fig. 2.11**. In order to reduce the effect of the corner-related coupling effect, all the 2×2 neighbored meta-atoms are grouped together as a macro unit. **Figure 2.11d** depicts the designed meta-atom, with a size of 15 mm × 15 mm × 5.2 mm, which is composed of three subwavelength-scale square metallic patches printed on a dielectric substrate (Rogers 3010) with dielectric constant of 10.2. Any two adjacent patches are connected via a PIN diode (BAR 65-02L), and each PIN diode has two operation states controlled by the applied bias voltage. Illuminated by an x-polarized plane wave, such a meta-atom supports

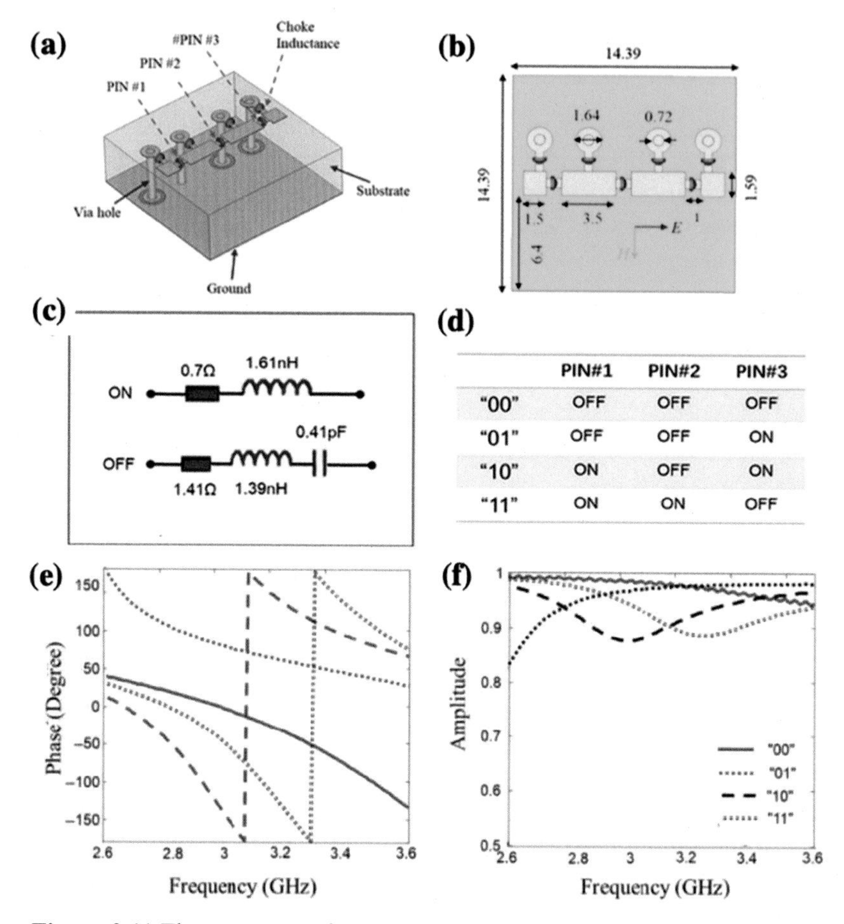

	PIN#1	PIN#2	PIN#3
"00"	OFF	OFF	OFF
"01"	OFF	OFF	ON
"10"	ON	OFF	ON
"11"	ON	ON	OFF

Figure 2.11 The meta-atom for realizing the two-bit reprogrammable coding metasurface. **(a)** Schematic figure and **(b)** top view of the metamaterial particle to realize the coding metasurface with detailed geometrical parameters. **(c)** The effective circuit models of the diode biased at the "ON" and "OFF" states. **(d)** The corresponding phase responses of the metamaterial particle as the diode is biased at the "OFF" and "ON" states over a range of frequencies. **(e)** Phase and **(f)** amplitude of the metasurface supercell. [16]

four different phase responses, denoted as "00" (state 0), "01" (state 1), "10" (state 2), "11" (state 3), and determined by controlling the ON/OFF states of the three PIN diodes in a suitable combination, corresponding to four digitized phase levels 0, $\pi/2$, π, and $3\pi/2$. It can be observed that the phase difference between two neighboring states falls in the range ($90° - 15°$, $90° + 15°$) around the frequency of 3.2 GHz. As a consequence, these meta-atoms can be treated as unit cells of a 2-bit digital coding metasurface, mimicking "00," "01," "10," and "11." It can be observed from **Fig. 2.11f** that the simulated reflection can reach a high efficiency above 85 percent for all digital states around 3.2 GHz. The diode is in the ON (or OFF) state with an applied external bias voltage of 3.3 V (or 0 V). To isolate the DC feeding port and microwave signal, three choke inductors of 30 nH are used in each metasurface atom. When the PIN diode is in the ON state, it is equivalent to a series circuit with parasitic inductance and resistance; while in the OFF state it is equivalent to a series circuit with parasitic inductance, capacitance, and resistance.

Now, we turn to discuss the optimal control coding patterns of the developed two-bit reprogrammable coding metasurface for generating the desired OAM beams, which can be obtained by modifying the modified GS algorithm discussed in Section 2.2. To that end, the source inversion technique is employed to calibrate the current distributions of $\mathbf{J}^{(00)}$, $\mathbf{J}^{(01)}$, $\mathbf{J}^{(10)}$, and $\mathbf{J}^{(11)}$, when the meta-atom at the states of "00," or "01," or "10," or "11" is illuminated by a plane wave. Then, the four-phase quantization is performed on the metasurface in each iteration step, in which careful consideration has been given to the back/forward propagation operations.

2.4.2 Results

Simulation and experimental results selected from [18] are provided here to show the performance of an above two-bit reprogrammable coding metasurface in generating the OAM beams with the controllable topological charges and vortex centers. In experiments, the vector network analyzer was used to acquire the response data by measuring the transmission coefficients (S_{21}). More specifically, a horn antenna and a standard waveguide tip are connected to two ports of VNA through two 4-m-long 50-Ω coaxial cables. The horn antenna is used to generate the x-polarized wave for normally illuminating the digital coding metasurface, and the waveguide tip mounted on a mechanical-movement scanning platform is used to acquire the response data on a near-field scanning plane with a resolution of 5 mm × 5 mm. Once the near-field distribution is achieved through a simple postprocess of spatial low-pass filtering, the technique of near-to-far field transformation is then employed to obtain the radiation distribution at an arbitrary observation distance. In addition, the distance between the transmitting antenna and the metasurface is 0.576 m. To suppress the measurement noise level, the average number and filter bandwidth in VNA are set to 10 and 10 kHz, respectively.

Case Study 1 Generating single-mode OAM beams

Figure 2.12 reports the experimental results to produce the origin-centered OAM-carrying beams with different topological charges of $\ell = 0, 1, 2, 3, 4, 5, 6$ when the metasurface is programmed with different coding patterns by FPGA. In the first column of **Fig. 2.12**, the corresponding coding patterns of the metasurface for generating the origin-centered OAM-carrying beams are plotted. The corresponding amplitude and phase distributions at the distance of 3 m away from the metasurface are presented in the second and third columns of **Fig. 2.12**, respectively. The spectrum decomposition of OAM is further performed, and the results are shown in the rightmost column of **Fig. 2.12**. In our calculation, for each OAM mode, the signal is extracted along the circle covered by the main lobe. From this set of results, it can be observed that the measured topological charge accounts for the highest proportion, being consistent with the desired results, whereas the other OAM components give rise to marginable contributions. The mismatch between the simulation and the experimental results is due to measurement errors, substrate material loss, and the imperfect quality of commercial PIN diodes.

To show the flexibility of the designed metasurface in generating the OAM-carrying beams with controllable vortex centers, we conduct a set of CST-based full-wave simulations. **Figure 2.13** shows the results of the OAM-carrying beams with different vortex centers generated by the designed metasurface. It can be immediately observed that the vortex centers of the OAM-carrying beams with different topologies can be dynamically controlled by using the designed two-bit reprogrammable coding metasurface, when the metasurface is reconfigured with corresponding control coding sequences. For instance, the OAM-carrying beams with topological charges $l = 1, 2, 3$ can be manipulated to steer toward $\theta = \pm 20°, 0°$, and $\pm 40°$ when the metasurface is controlled by changing its digital coding sequence.

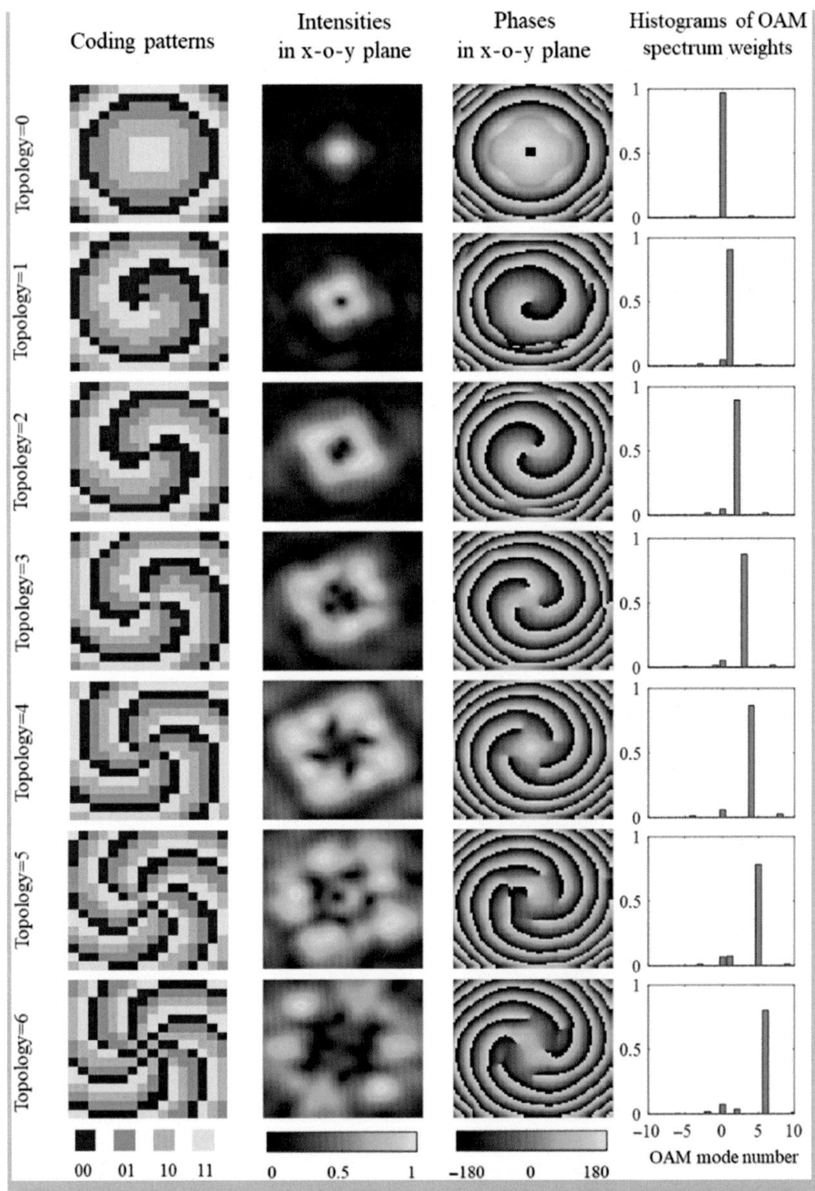

Figure 2.12 The measured results of single-mode OAM-carrying beams with different topological charges of $l = 0, 1, 2, 3, 4, 5$, and 6. The first column: seven control coding patterns of the programmable metasurface. The second column: the corresponding distributions of field intensities in the x-o-y plane at the distance of $z = 3$ m away from the metasurface. The third column: the corresponding phase distributions in the x-o-y plane at the distance of $z = 3$ m. The fourth column: the corresponding histograms of OAM spectrum weights.[18]

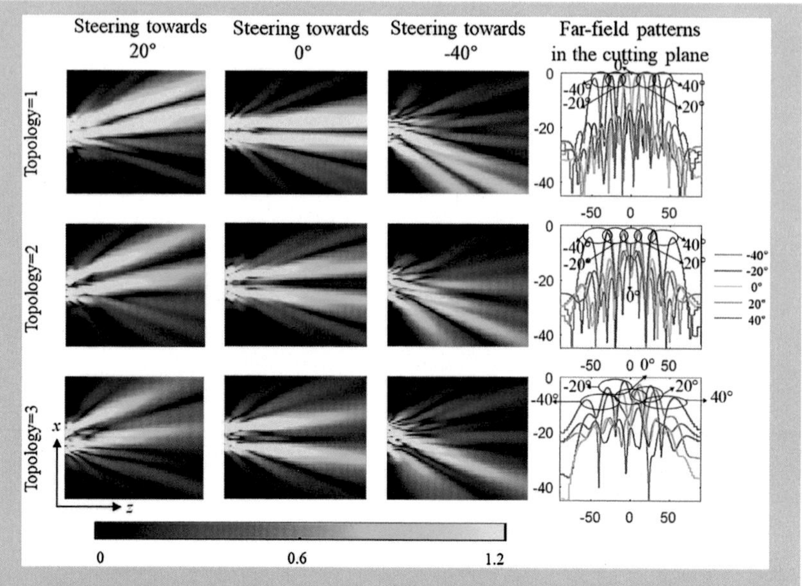

Figure 2.13 Simulation results of the single-mode OAM beams with different vortex centers and topological charges of 1, 2, and 3, which are generated by the designed two-bit reprogrammable coding metasurface.

CASE STUDY 2 GENERATING MULTIMODE OAM BEAMS

From the preceding discussion, we clearly show that the designed metasurface can be used to dynamically control the topological charges and vortex centers of the *single-mode* OAM-carrying beams. Here, we would like to experimentally demonstrate that our metasurface can also be used to dynamically generate and control *multimode* OAM-carrying beams. By "multimode OAM beam" we mean that the generated OAM beam has multiple vortex centers or/and topological charges, and these modes can be independently controlled. **Figure 2.14** shows some selected experimental results of generated multimode OAM-carrying beams with controllable vortex centers. The dual-mode OAM beams with topological charges $l = -1$ and $l = 1$ are designed to respectively steer to $\theta = 30°$ and $\theta = 20°$ in the y-o-z plane, as shown in the first row in **Fig. 2.14**. The second row in **Fig. 2.14** illustrates the dual-mode OAM beams with topological charges $l = 2$ and $l = -1$, which are manipulated to respectively steer to $\theta = 10°$ and $\theta = 25°$ in the x-o-z plane. Also, the triple-mode and four-mode OAM-carrying beams with different topological charges and vortex centers in both the y-o-z and x-o-z planes have been depicted in the third and fourth rows of **Fig. 2.14**. It can be safely concluded that not only the vortex center but also the mode number of

OAM-carrying beams can be dynamically and flexibly controlled by using the designed two-bit reprogrammable coding metasurface.

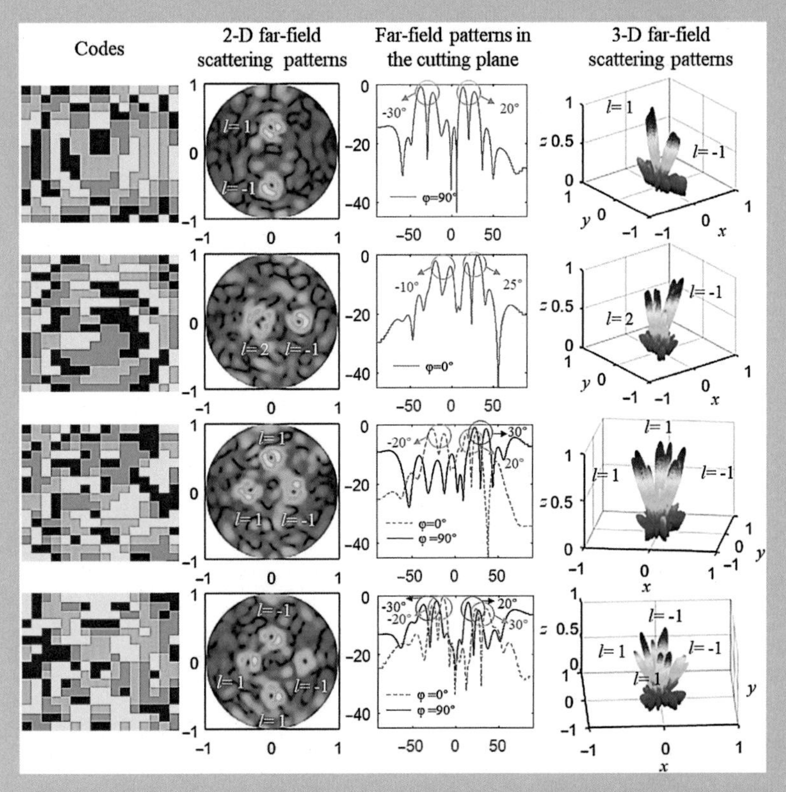

Figure 2.14 Results of multi-mode OAM beams with different vortex centers and topological charges. The first column: Coding patterns of the designed programmable metasurface. The second column: Corresponding 2D far-field radiation patterns. The third column: Corresponding far-field patterns in the cutting planes of φ = 0 and φ = 90°. The fourth column: Corresponding 3D far-field scattering patterns.

Case Study 3 Generating focused OAM beams

It can be observed from the preceding results that the OAM beam suffers from an important drawback: the OAM beam will inevitably diverge with the growth of propagation distance, especially for the high-order cases. As shown in **Fig. 2.12**, the divergence angle of the OAM beam increases with the growth of scanning angle and the gain of the two main lobes not only declines but also becomes obviously unbalanced when the OAM beams steer toward a larger angle. And this situation becomes much worse when the topological charge is enlarged. To address these issues, Shuang et al. presented an effective and robust strategy for generating focused OAM beams, which was motivated by the so-called convolutional theorem developed in [44]. In our specific implementations, we performed the module-4 additive operation on two coding patterns of the two-bit reprogrammable coding metasurface, where one coding pattern is for generating an *l*-order OAM-carrying beam, and the other coding pattern is for generating a well-focused spot at a desired location. With such a simple strategy, we can easily obtain the focused OAM-carrying beam. **Figure 2.15** compares the profiles of OAM-carrying beams in the x-o-y plane at $z = 1$ m with and without the use of the focusing strategy. It is clear that the waist widths of the focused OAM-carrying beams have notably shrunk compared to the unfocused OAM beams. Now, we conclude that the designed two-bit reprogrammable coding metasurface is capable of dynamically generating the high-order OAM-carrying beams with the remarkably improved FWHMs.

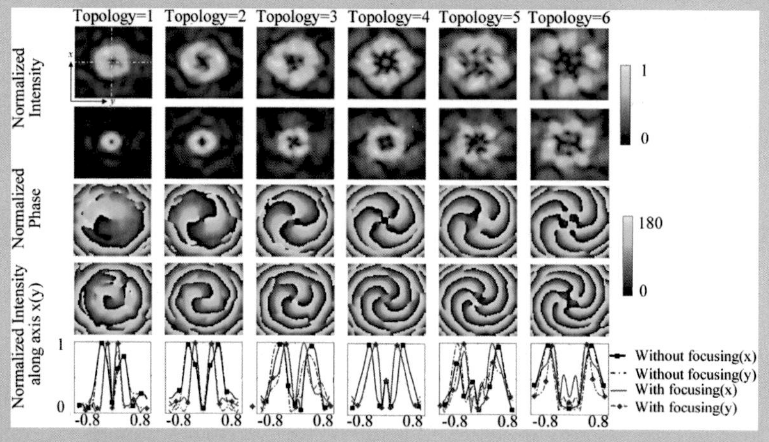

Figure 2.15 Comparisons of the single-mode OAM beams with different topological charges in the x-o-y plane at $z = 1$ m with and without using the focusing strategy. The first row: Normalized intensities of the OAM beams without using the focusing technique. The second row: Normalized intensities of the OAM beams using the focusing method. The third row: Phase distributions of the OAM beams in the x-o-y plane at $z = 1$ m corresponding to the first row. The fourth row: Phase distributions of the OAM beams in the x-o-y plane at $z = 1$ m corresponding to the second column. The fifth row: Profiles of the normalized intensities along the white cutting lines.

3 Information Fundamentals of Reprogrammable Metasurfaces

In this section we give theoretical insights into the performance of one-bit information-encoding metasurfaces [45]. We provide the signal model of a reprogrammable coding metasurface as a *wireless* multiport device that links various incident beams to various outgoing beams, where the analytical representations of system responses in various cases (single-input single-output, SISO; single-input multiple-output, SIMO; and multiple-input multiple-output, MIMO) are discussed. Afterwards, we investigate the information capacity of a one-bit coding metasurface. Our findings indicate that the one-bit coding metasurface can have a satisfactory performance comparable to that of continuously tunable metasurfaces in many practical scenarios.

3.1 Signal Representation of One-Bit Coding Metasurfaces

To date, many practical realizations of the reprogrammable metasurfaces rely on one-bit (or few-bit) coding in order to further reduce the system cost, complexity, and energy consumption. A one-bit meta-atom is controlled with an externally applied one-bit voltage and correspondingly has two distinct response states denoted by "0" and "1." For instance, upon illumination with a plane wave, the state "0" ("1") may correspond to a reflection phase of 0° (180°). Given the huge success of one-bit coding metasurfaces in diverse applications, a fundamental question naturally arises [46–48]: what is the achievable information capacity of a programmable metasurface built with one-bit coding meta-atoms ($R = \pm 1$) compared to the ideal case of continuously tunable meta-atoms ($R = A\exp(j\varphi)$ with $|A| \leq 1$ and $0 \leq \varphi < 2\pi$)? To address the quantization question for information-encoding metasurfaces operating in free space, we give a mathematical representation of the system response in the case of one-bit coding for three representative scenarios: SISO, SIMO, and MIMO. Our analysis reveals that the one-bit quantization can give rise to undesirable effects like energy leakage, parasitic unwanted beams, and channel cross-talk. We derive analytical expressions for the signal interference and reduction in channel capacity, and find that the performance deterioration is negligible and can be limited under mild constraint relaxations in many practical applications.

Some notations and assumptions used throughout this section are listed as follows: (1) The macro meta-atom is referred to as a set of adjacent meta-atoms, and the macro meta-atoms have mutually independent EM responses. (2) The entire metasurface is composed of $M \times N$ macro meta-atoms. Throughout this section we consider a 24×24 array of programmable meta-atoms, each of size

54×54 mm^2. (3) The meta-atom has the EM responses as $R = A\exp(j\varphi)$. (4) The EM response of macro meta-atom is independent of the angle of incidence as well as inter-element coupling effects. Nonetheless, we have previously used this specific model successfully to configure a programmable metasurface prototype in imaging [49] and communication [50] applications such that the model is known to be a good approximation of the experimental reality.

Case I: SISO

For SISO, the one-bit coding metasurface is used to establish a wireless channel linking a source at $\boldsymbol{r}_s$ with an intended receiver at $\boldsymbol{r}_q$, as shown in **Fig. 3.1(a)**. With the loss of generality, the signal power level of the source is assumed to be $\mathcal{P}$ and the signal is desired to acquire a phase ϕ_q ($0 \leq \phi_q < 2\pi$) when it reaches $\boldsymbol{r}_q$ in a phase shift keying (PSK) scheme. We emphasize that the information is *not* encoded by the source at $\boldsymbol{r}_s$ but through the EM wave's modulation by the metasurface. However, the scenario of information being encoded by the source is a special case of our model, where only the intensity of the radiation beam is concerned. A closed-form estimate of a suitable one-bit coding pattern of the metasurface for this purpose is

$$C_{m,n}^{SISO} = \text{sign}\left[\cos\left(\widetilde{\phi}_{nm}^{SISO}\right)\right]. \tag{3.1}$$

where $\widetilde{\phi}_{nm}^{SISO} \equiv \widetilde{\phi}_{m,n}\left(\boldsymbol{r}_q; \boldsymbol{r}_s\right)$,

$$\widetilde{\phi}_{m,n}\left(\boldsymbol{r}_q; \boldsymbol{r}_s\right) = \Delta_{m,n}\left(\boldsymbol{r}_q; \boldsymbol{r}_s\right) + \phi\left(\boldsymbol{r}_q\right),$$

$$\Delta_{m,n}\left(\boldsymbol{r}_q; \boldsymbol{r}_s\right) = k\left(\underbrace{\left|\boldsymbol{r}_s - \boldsymbol{r}_{m,n}\right|}_{R_{nm}(r_s)} + \underbrace{\left|\boldsymbol{r}_q - \boldsymbol{r}_{m,n}\right|}_{R_{nm}\left(r_q\right)}\right).$$

Herein, $\phi_q \equiv \phi\left(\boldsymbol{r}_q\right)$ ($0 \leq \phi_q < 2\pi$) denotes the intended phase directed from the source to the receiver at $\boldsymbol{r}_q$ through the metasurface. Considering that the location of the source $\boldsymbol{r}_s$ is usually inaccurate or even unknown, we assume that the true source location is $\mathbf{r}'$. Then, the response at $\mathbf{r}$ can be derived as (see Appendix 1):

$$\hat{\mathcal{H}}_{\text{SISO}}(\mathbf{r}, \mathbf{r}'; q, s) = \underbrace{E_1^{SISO}(\boldsymbol{r}, \boldsymbol{r}'; q, s)}_{\text{leading term}} + \underbrace{\sum\nolimits_{p=-\infty, p\neq 1}^{\infty} B_p^{SISO} A_p^{SISO}(\boldsymbol{r}, \boldsymbol{r}'; q, s)\exp\left(jp\phi_q\right)}_{\text{perturbation terms}} \tag{3.2}$$

Herein, j is the imaginary unit and

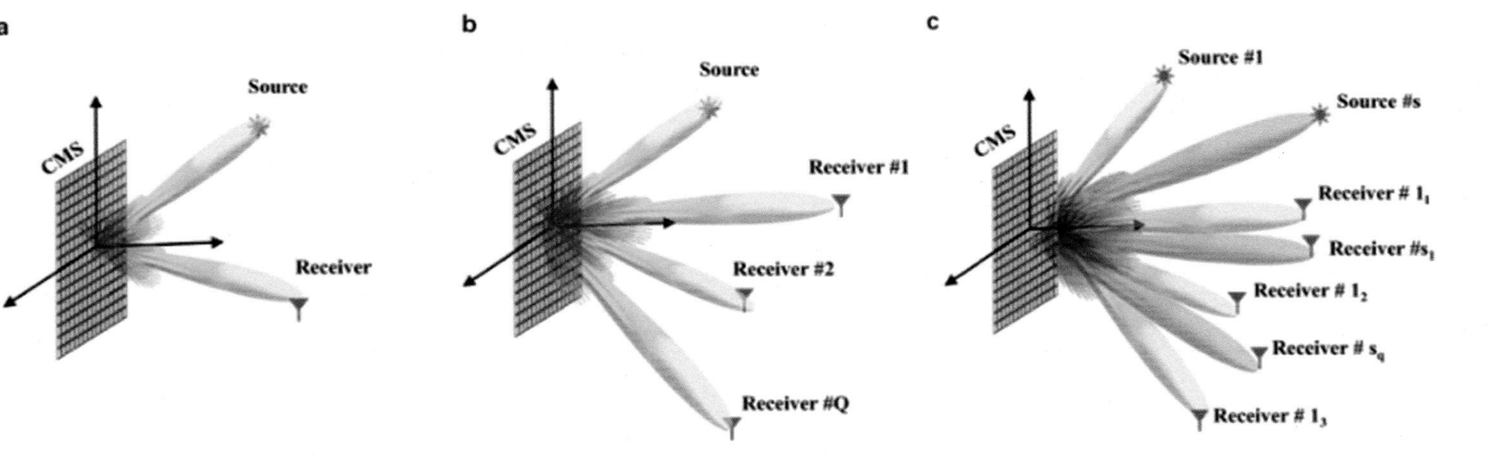

Figure 3.1 Illustration of how a programmable coding-metasurface (CMS) acts as a wireless multiple-port device. **(a)** SISO: CMS serves as a wireless ultrathin device with one input port and one output port, linking one source to one intended receiver. **(b)** SIMO: CMS serves as a single-input multiple-output wireless device, connecting one source with Q ($Q > 1$) intended receivers (Receiver #1, Receiver #2, . . ., Receiver # Q). **(c)** MIMO: CMS serves as a wireless ultrathin device with S input ports and $\sum_{i=1}^{S} Q_i$ output ports. Each source (e.g., Source # s, $s = 1, 2, \ldots, S$) is linked to Q_i intended receivers (i.e., Receiver # s_1, Receiver # s_2, . . ., Receiver # s_{Q_s}) by CMS.

$$E_p^{SISO}(\boldsymbol{r},\boldsymbol{r}';q,s) = B_p^{SISO} A_p^{SISO}(\boldsymbol{r},\boldsymbol{r}';q,s)\exp\left(j\phi_q\right),$$

$$A_p^{SISO}(\boldsymbol{r},\boldsymbol{r}';q,s) = \sum\nolimits_{m,n} \frac{\exp\left[j\left(p\Delta_{nm}\left(\boldsymbol{r}_q;\boldsymbol{r}_s\right) - \Delta_{nm}(\boldsymbol{r};\boldsymbol{r}')\right)\right]}{R_{nm}(\boldsymbol{r})R_{nm}(\boldsymbol{r}')},$$

$$\text{and } B_p^{SISO} = \begin{cases} -\dfrac{j^{p+1}}{\pi}\dfrac{2}{p}, \text{ if } p \text{ is odd} \\ 0, \quad \text{else} \end{cases}.$$

Note that $|B_1^{SISO}| = |B_{-1}^{SISO}|$ and $R_{nm}(\boldsymbol{r}) = |\boldsymbol{r} - \boldsymbol{r}_{nm}|$, where $\boldsymbol{r}_{nm}$ denotes the location of the (m, n) meta-atom. **Equation (3.2)** offers an important insight. In particular, the leading term E_1^{SISO} represents the system response of continuous metasurface but corrected by a multiplicative factor of $B_1^{SISO} = \frac{2}{\pi} \sim 0.64$ (corresponding to about 3 dB energy loss). Meanwhile, the perturbation terms characterize the unwanted parasitic beams that divert the energy from the source into directions other than that of the intended receiver due to the one-bit quantization of the meta-atom programmability.

The spatial maps of $|\hat{\mathcal{H}}_{SISO}|$ for representative cases are plotted in **Fig. 3.2** in order to illustrate the roles of different terms. Firstly, we consider two cases for which the source and the intended receiver are in the near field of the metasurface. For a receiver on or off the 0° azimuth, **Fig. 3.2a and b** respectively show plots of $|\hat{\mathcal{H}}_{SISO}|$ both with continuous and one-bit coding metasurfaces, where the first few nonzero terms $|E_p^{SISO}|$ are visualized. We observe that the leading term $p = 1$ dominates the system response $|\hat{\mathcal{H}}_{SISO}|$ and there is no significant difference between the continuous coding and one-bit coding. While the focus is about 3 dB weaker with the one-bit coding, the quantization energy loss is statistically uniformly distributed over the entire space. Overall, in these two cases the performance of one-bit coding is of comparable quality to that with the continuous coding. Next, we consider two similar scenarios in **Fig. 3.2c and d** for which both the source and the intended receiver are in the far field region. Here, the perturbation terms ($p \neq 1$) have nonnegligible contributions to the system responses $|\hat{\mathcal{H}}_{SISO}|$ but can be truncated at $P = 3$ (RMSE < 0.5%). Unless the receiver is at the 0° azimuth, multiple unwanted parasitic beams appear to interfere with the desired beam. In particular, the $p = -1$ term appears to be mirror-symmetrical with respect to the normal direction $\hat{\boldsymbol{n}}$ of the metasurface.

To shed more light on these far-field limitations, we simplify the E_p^{SISO} terms under a far-field approximation:

$$E_p^{SISO}(\boldsymbol{r},\boldsymbol{r}_S;q,s) \approx B_p^{SISO} MN \frac{\exp\left[jk\left(pr_q - r + (p-1)r_s\right)\right]}{rr_s}\exp\left(jp\phi_q\right),$$

for

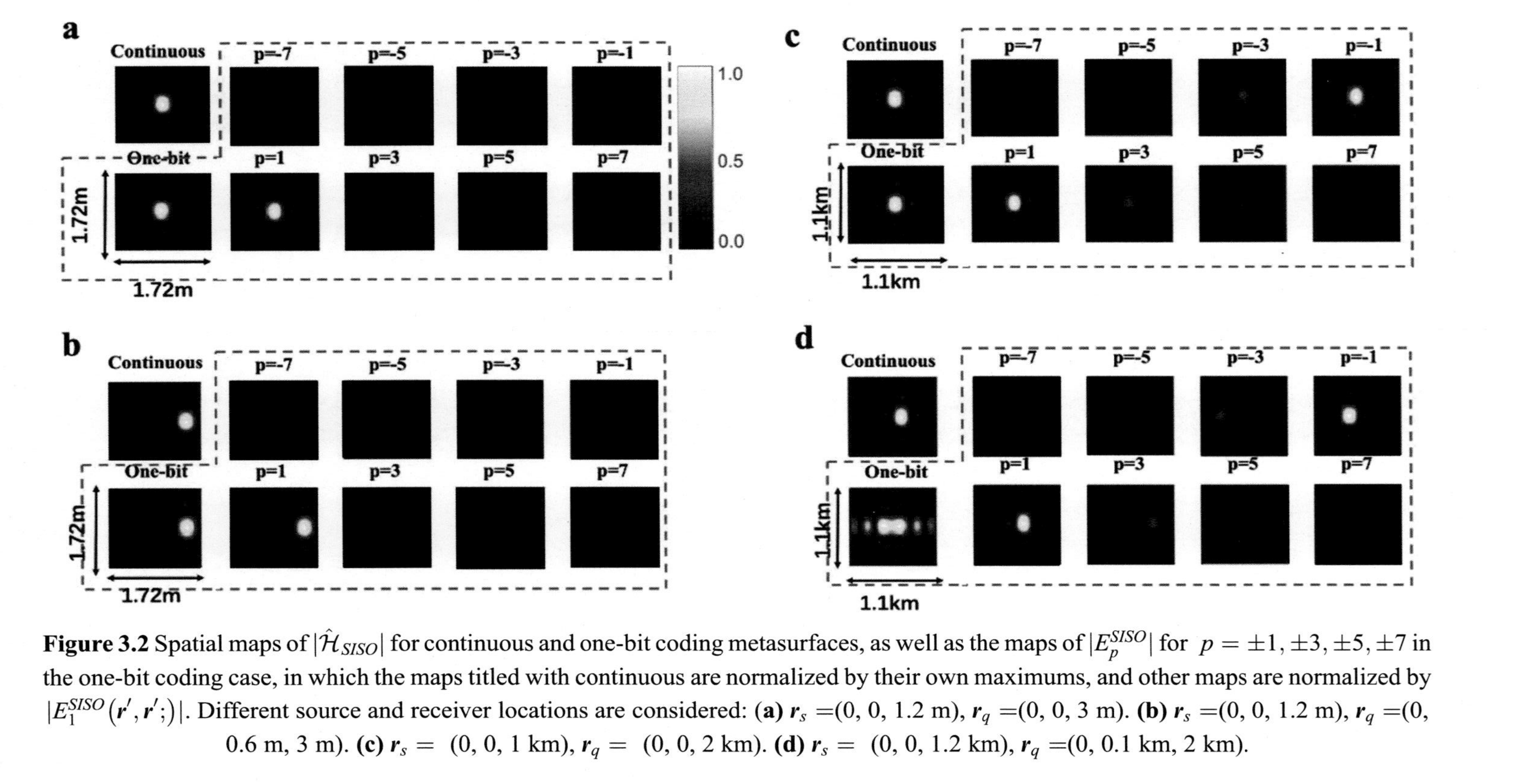

Figure 3.2 Spatial maps of $|\hat{\mathcal{H}}_{SISO}|$ for continuous and one-bit coding metasurfaces, as well as the maps of $|E_p^{SISO}|$ for $p = \pm 1, \pm 3, \pm 5, \pm 7$ in the one-bit coding case, in which the maps titled with continuous are normalized by their own maximums, and other maps are normalized by $|E_1^{SISO}(\boldsymbol{r}', \boldsymbol{r}';)|$. Different source and receiver locations are considered: **(a)** $\boldsymbol{r}_s$ =(0, 0, 1.2 m), $\boldsymbol{r}_q$ =(0, 0, 3 m). **(b)** $\boldsymbol{r}_s$ =(0, 0, 1.2 m), $\boldsymbol{r}_q$ =(0, 0.6 m, 3 m). **(c)** $\boldsymbol{r}_s =$ (0, 0, 1 km), $\boldsymbol{r}_q =$ (0, 0, 2 km). **(d)** $\boldsymbol{r}_s =$ (0, 0, 1.2 km), $\boldsymbol{r}_q$ =(0, 0.1 km, 2 km).

$$\hat{\boldsymbol{r}} - p\hat{\boldsymbol{r}}_q + (1-p)\hat{\boldsymbol{r}}_s \parallel \hat{\boldsymbol{n}}. \tag{3.3}$$

Then the desired beam ($p = 1$) in the far-field approximation is

$$E_1^{SISO}(\boldsymbol{r}, \boldsymbol{r}_S; q, s) \approx B_1^{SISO} MN \frac{\exp\left[jk\left(r_q - r\right)\right]}{rr_s} exp\left(j\phi_q\right), \text{ for } \hat{\boldsymbol{r}} = \hat{\boldsymbol{r}}_q \tag{3.4}$$

and there are multiple significant unwanted parasitic beams corresponding to the values of $p \neq 1$. In particular, the expression for the $p = -1$ term in the far-field region is written as

$$E_{-1}^{SISO}(\boldsymbol{r}, \boldsymbol{r}_S; q, s) \approx B_{-1}^{SISO} MN \frac{\exp\left[-jk\left(r_q + r - 2r_s\right)\right]}{rr_s} \exp\left(-j\phi_q\right),$$
$$\text{for } \hat{\boldsymbol{r}}_q + \hat{\boldsymbol{r}} + 2\hat{\boldsymbol{r}}_s \parallel \hat{\boldsymbol{n}}. \tag{3.5}$$

Here, $|B_1^{SISO}| = |B_{-1}^{SISO}|$ confirms the previous observation in **Fig. 3.2(d)** of mirror-symmetry to the desired beam in terms of amplitude and now it is 180° out of phase with the desired beam.

Now, we utilize **Eq. (3.2)** to study the information-encoding capabilities of the one-bit coding metasurface in wireless communications with M_q-level PSK, which can ideally achieve the phase quantization of $\hat{\mathcal{H}}_{\text{SISO}}$ as $\phi_q \in \left\{\frac{2\pi(i-1)}{M_q},\ i = 1, 2, \ldots, M_q\right\}$. To quantify to what extent the one-bit coding metasurface is capable of approaching these desired values, we define the following metric for the achievable phase resolution:

$$\Delta\varphi_{SISO}\left(\boldsymbol{r}_q\right) = \min_{i,j}\left|\varphi_{SISO}^{(i)}\left(\boldsymbol{r}_q\right) - \varphi_{SISO}^{(j)}\left(\boldsymbol{r}_q\right)\right|, \tag{3.6}$$

where $\varphi_{SISO}^{(i)}$ is the phase of $\hat{\mathcal{H}}_{\text{SISO}}$ for the ith intended phase. To quantify the difference between the performance with one-bit and ideal coding, we define the following metric:

$$EC^{SISO} = \frac{1}{MN}\sum_{m,n}\left|\text{sign}\left[\cos\left(\tilde{\phi}_{nm}^{SISO}\right)\right] - \cos\left(\tilde{\phi}_{nm}^{SISO}\right)\right|. \tag{3.7}$$

Figures 3.3a and e plot the dependence of $\Delta\varphi_{SISO}\left(\boldsymbol{r}_q\right)$ and EC^{SISO}, respectively, on $\boldsymbol{r}_q$ for QPSK ($M_q = 4$) with a normally incident source deployed at a distance of 2 km from the metasurface, namely in its far-field. The ability of the one-bit coding metasurface to perform QPSK information encoding with high fidelity is immediately obvious, except for the case when the receiver is on the azimuth 0° and in the far-field region. This result is in line with our abovementioned interpretation of **Eq. (3.6)**. In a scenario with a source incident at an oblique angle, as shown in **Fig. 3.3d and f**, there appears to be a similar limitation if $\hat{\boldsymbol{r}}_q + \hat{\boldsymbol{r}}_s \parallel \hat{\boldsymbol{n}}$. Indeed, in this case the far-field approximation of $\tilde{\phi}_{nm}^{SISO}$ reads

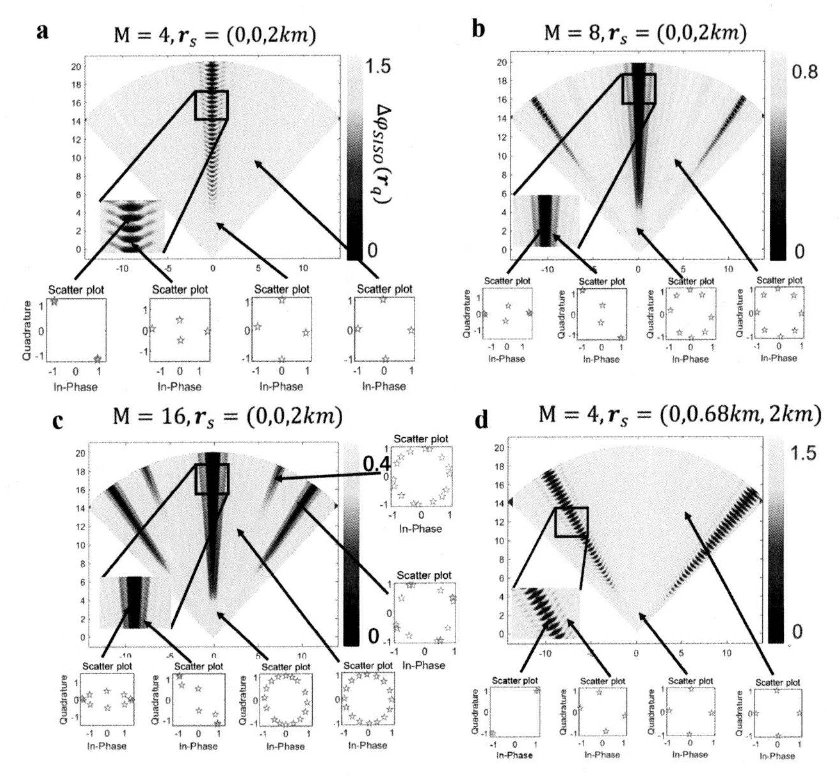

Figure 3.3 M-level PSK with a one-bit coding programmable metasurface in SISO. **(a–d)** Dependence of phase resolution $\Delta\varphi_{SISO}$ on the receiver's location $\boldsymbol{r}_q$. Considered receiver locations are within a distance between 0 and 20 m from the metasurface for an azimuth between –45° and 45°. The insets provide representative constellation diagrams. Different levels of PSK and source location are considered, as indicated in the subfigure titles. **(e–h)** Dependence of EC^{SISO} on $\boldsymbol{r}_q$ for the same settings as in **a–d**. In addition, the metasurface has been marked with a yellow solid rectangle.

$$\begin{aligned}\tilde{\phi}_{nm}^{SISO} &= k(|\boldsymbol{r}_s - \boldsymbol{r}_{m,n}| + |\boldsymbol{r}_q - \boldsymbol{r}_{m,n}|) + \phi_q \\ &\approx k\Big(r_s + r_q - \boldsymbol{r}_{m,n} \cdot \big(\hat{\boldsymbol{r}}_s + \hat{\boldsymbol{r}}_q\big)\Big) + \phi_q \\ &\approx k\big(r_s + r_q\big) + \phi_q \quad \text{for } \hat{\boldsymbol{r}}_q + \hat{\boldsymbol{r}}_s \parallel \hat{\boldsymbol{n}},\end{aligned} \tag{3.8}$$

such that the control coding pattern of the one-bit coding programmable metasurface is $C_{m,n}^{SISO} = \text{sign}\Big[\cos\Big(k\big(r_s + r_q\big) + \phi_q\Big)\Big]$. Again, it is obvious why a receiver in the far-field with $\hat{\boldsymbol{r}}_q + \hat{\boldsymbol{r}}_s \parallel \hat{\boldsymbol{n}}$ cannot distinguish M_q-level PSK information if $M_q > 2$. In such cases, PSK must be limited to $M_q = 2$

(BPSK). The pronounced increase of EC^{SISO} for this scenario in **Fig. 3.3e–h** further confirms that this limitation is attributed to the one-bit coding of the programmable metasurface. Higher-order PSKs with $M_q = 8$ and $M_q = 16$ are considered in **Fig. 3.3b–c** and **Fig. 3.3f–g** in terms of $\Delta\varphi_{SISO}(\boldsymbol{r}_q)$ and EC^{SISO}, respectively. In light of the well-known reciprocal property, we conclude that the one-bit coding metasurface is capable of efficiently manipulating the EM information at least up to 16-level PSK in the considered setup, if the source and the receiver are deployed in the near-field region. In general, efficient manipulation of EM information becomes increasingly challenging as M_q increases due to the increasing importance of unwanted parasitic beams.

Case II: SIMO

We first consider the SIMO setting, in which a single source at $\boldsymbol{r}_s$ is linked to Q ($Q > 1$) receivers via directive radiation beams, upon interacting with the metasurface. The qth beam is aimed at the intended receiver at $\boldsymbol{r}_q$ with desired phase ϕ_q ($0 \le \phi_q < 2\pi$) for the PSK information encoding. For this purpose, the closed-loop estimate of the required one-bit control coding pattern of the metasurface is

$$C_{m,n}^{SIMO} = \text{sign}\left[\sum\nolimits_{q=1}^{Q} \cos\left(\tilde{\phi}_{nm}^{SIMO}(q)\right)\right], \tag{3.9}$$

where $\tilde{\phi}_{nm}^{SIMO}(q) \equiv \tilde{\phi}_{m,n}(\boldsymbol{r}_q; \boldsymbol{r}_s)$,

$$\tilde{\phi}_{m,n}(\boldsymbol{r}_q; \boldsymbol{r}_s) = \Delta_{m,n}(\boldsymbol{r}_q; \boldsymbol{r}_s) + \phi_q,$$

and $\Delta_{m,n}(\boldsymbol{r}_q; \boldsymbol{r}_s) = k\left(\underbrace{|\boldsymbol{r}_s - \boldsymbol{r}_{m,n}|}_{R_{nm}(r_s)} + \underbrace{|\boldsymbol{r}_q - \boldsymbol{r}_{m,n}|}_{R_{nm}(r_q)}\right).$

Following similar steps as those in SISO, we obtain the system response of the one-bit coding metasurface at $\boldsymbol{r}$ as (see **Appendix 2**):

$$\hat{\mathcal{H}}_{SIMO}(\boldsymbol{r}, \boldsymbol{r}'; \{q\}, s) = \underbrace{E_1^{SIMO}(\boldsymbol{r}, \boldsymbol{r}'; \{q\}, s)}_{leading\ term} +$$

$$\underbrace{\sum \{p_q\}/\{\sum q\ |p_q| = 1\ \&\ p_q \ne -1\} E_{\{p_q\}}^{SIMO}(\boldsymbol{r}, \boldsymbol{r}'; \{q\}, s) exp\left[j\sum\nolimits_{q=1}^{Q} p_q \phi_q\right]}_{perturbation\ terms}. \tag{3.10}$$

Here, $\{q\}$ denotes a collection of Q receivers at different locations, and $\{p_q\}$ is a collection of Q harmonic orders which comes from the aforementioned Q receivers one by one. In **Eq. (3.10)**,

$$E_1^{SIMO}(\boldsymbol{r},\boldsymbol{r}';\{q\},s) = B_1^{SIMO}\sum_{q=1}^{Q}\exp\left(j\phi_q\right)A_1^{SISO}(\boldsymbol{r},\boldsymbol{r}';q,s),$$

$$E_{\{p_q\}}^{SIMO}(\boldsymbol{r},\boldsymbol{r}';\{q\},s) = B_{\{p_q\}}^{SIMO}A_{\{p_q\}}^{SIMO}(\boldsymbol{r},\boldsymbol{r}';\{q\},s),$$

where $B_{\{p_q\}}^{SIMO} = -\frac{j}{\pi}j^{\sum_{q=1}^{Q}p_q}\int_{-\infty}^{\infty}\frac{1}{\xi}\prod_{q=1}^{Q}J_{p_q}(\xi)d\xi,$

$$B_0^{SIMO} = \frac{1}{\pi}\int_{-\infty}^{\infty}\frac{1}{\xi}J_1(\xi)J_0^{Q-1}(\xi)d\xi,$$

and

$$A_{\{p_q\}}^{SIMO}(\boldsymbol{r},\boldsymbol{r}';\{q\},s) = \sum_{m,n}\frac{\exp\left(j\left[\sum_{q=1}^{Q}p_q\Delta_{m,n}(\boldsymbol{r}_q;\boldsymbol{r}_s) - \Delta_{m,n}(\boldsymbol{r};\boldsymbol{r}')\right]\right)}{R_{nm}(\boldsymbol{r})R_{nm}(\boldsymbol{r}')}.$$

Note that $B_{\{p_q\}}^{SIMO} = 0$ when $\sum_{q=1}^{Q}|p_q|$ is an even number. Here, the Q-fold multiple summation of the second term in **Eq. (3.10)** can be approximated with an accuracy of RMSE < 0.5% under the condition of $\sum_q|p_q|\leq 3$.

Similar to the case of SISO, the leading term E_1^{SIMO} in **Eq. (3.10)** describes the system response of the corresponding continuous metasurface but with a multiplicative factor of B_1^{SIMO}. The collection $\{p_q\}/\{\sum_q|p_q| = 1\,\&\,p_q \neq -1\}$ describes the energy leakages due to the one-bit quantization in comparison to the continuous coding metasurface. Moreover, signal interferences and cross talks between different intended receivers arise due to the one-bit coding, which will be studied in the subsequent section.

Now, we utilize **Eq. (3.10)** to study the signal interference of SIMO by considering the application of metasurface-generated holograms. Specifically, we examine the capability of the one-bit coding metasurface to realize the holographic image of English letter "K." First, we assume that the desired phases over the profile of the letter "K" are randomly distributed, as shown in **Fig. 3.4a**. The normalized amplitude and phase of the achieved holographic image at a distance of 3 m away from the one-bit coding metasurface is shown in the top row of **Fig. 3.4c**, and the corresponding control coding pattern of the one-bit coding metasurface is reported. For comparison, the corresponding results obtained by the modified GS algorithm with and without the phase constraint are shown in the middle and bottom rows of **Fig. 3.4c**. In both cases we initialize the GS algorithm with $C_{m,n}^{SIMO}$. The convergence behavior of the GS algorithm with and without the phase constraint is plotted in **Fig. 3.4b**.

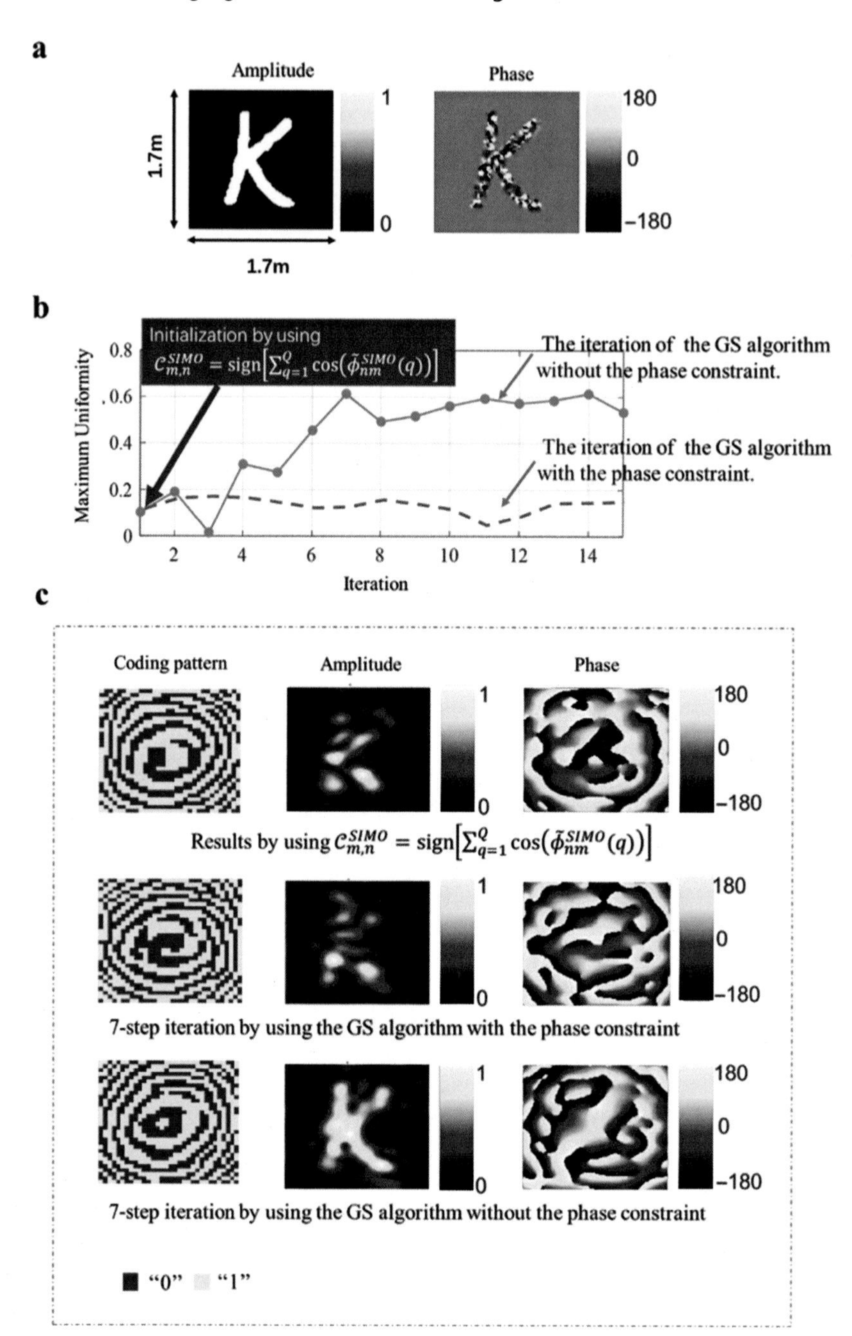

Figure 3.4 SIMO results for holographic imaging. The source is deployed at (0,0,2 m) and the observation plane is at a distance of 3 m from the one-bit coding metasurface. **(a)** Amplitude and phase of the targeted hologram to be generated by the one-bit coding metasurface. **(b)** Convergence of the GS algorithm with and without the phase constraint in terms of the maximum

Our results show that the proposed closed-form formula for obtaining the coding pattern of the one-bit coding metasurface, $C_{m,n}^{SIMO}$, gives satisfactory results with comparable quality to that achieved by the iterative GS algorithm. In both cases, the quantization-induced signal interference among different channels notably limits the quality of the holographic image. If the constraints are relaxed by ignoring the desired phases, the one-bit coding metasurface yields a holographic image of acceptable quality. A further possibility is to simplify the phase constraint by targeting a constant value for the entire letter "K." The corresponding results in **Fig. 3.5** for different targeted phase values reveal a setting, in which the one-bit coding metasurface produces acceptable holographic images in terms of both phase and amplitude. Then, it can be verified that the phase $\{\phi_q\}$ could provide controllable parameters to minimize the signal interference, as pointed out previously with respect to the phase bias $\{\theta_i\}$.

Case III: MIMO

We provide the MIMO system response of a one-bit coding programmable metasurface. In the MIMO setting, the metasurface is illuminated by S independent sources, and the s th source generates Q_s independent radiation beams upon interaction with the metasurface. For the s th source located at $\boldsymbol{r}_s(s = 1, 2, \ldots, S)$, the q_s th beam is aimed at the q_s th intended receiver located at $\boldsymbol{r}_{q_s}(q_s = 1, 2, \ldots, Q_s)$ with a desired phase $\phi_{q_s}(0 \leq \phi_{q_s} < 2\pi)$ for PSK information encoding. For this purpose, the closed-loop estimate of the required one-bit coding pattern of the metasurface is

$$C_{m,n}^{MIMO} = \text{sign}\left[\sum_{s=1}^{s=S} \sum_{q_s=1}^{q_s=Q_s} \cos\left(\tilde{\phi}_{nm}^{MIMO}(q_s)\right)\right], \tag{3.11}$$

where $\tilde{\phi}_{nm}^{MIMO} \equiv \tilde{\phi}_{m,n}\left(\boldsymbol{r}_q; \boldsymbol{r}_s\right)$,

---

Caption for Figure 3.4 (cont.)

uniformity as the figure of merit. In both cases, the GS algorithm is initialized with $C_{m,n}^{SIMO} = \text{sign}\left[\sum_{q=1}^{Q} \cos\left(\tilde{\phi}_{nm}^{SIMO}(q)\right)\right]$. (**c**) Results using $C_{m,n}^{SIMO}$ are shown in the top row, where the coding pattern of the one-bit metasurface, the amplitude, and the phase of the holographic image are shown on the left, middle, and right, respectively. The corresponding results using the GS algorithm with and without the phase constraint are shown in the middle and bottom rows, respectively.

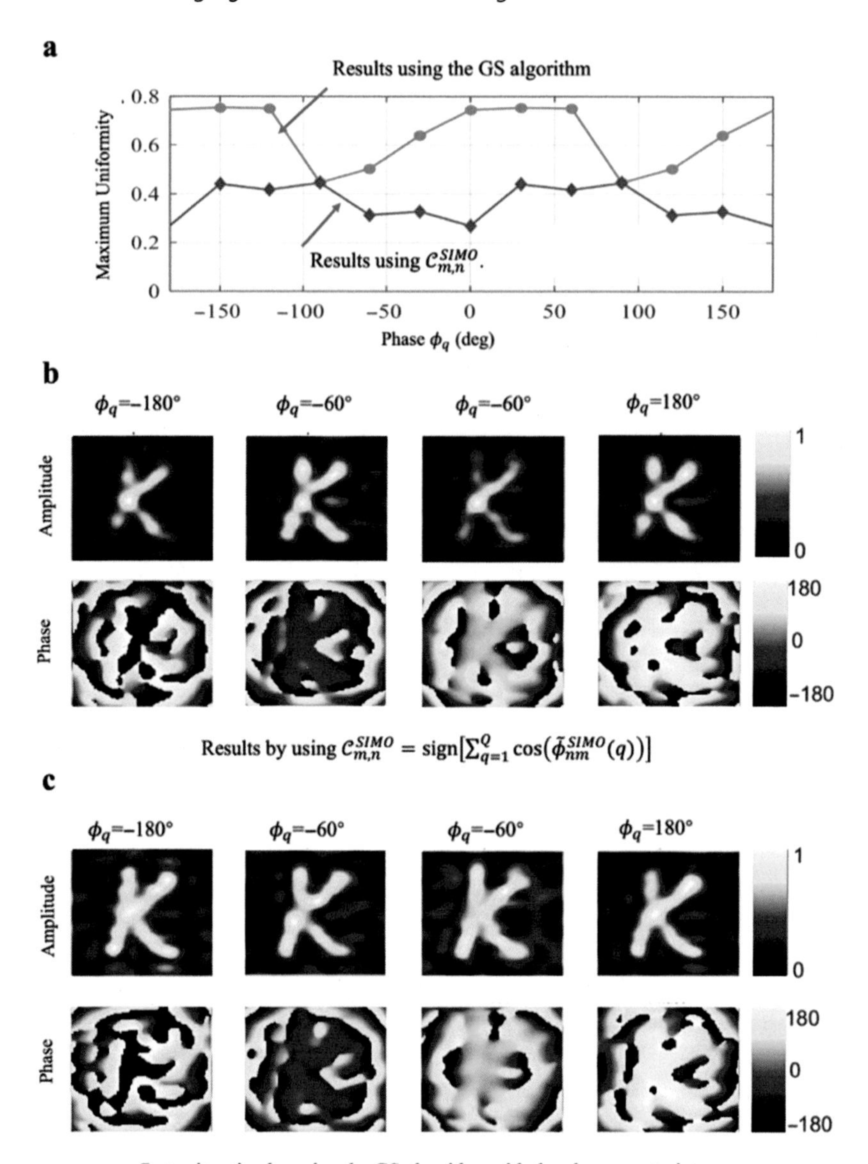

Figure 3.5 Relaxing phase constraints in the SIMO holography. The setup is the same as that in **Fig. 3.4** except that the phase of the targeted holographic image is set to be constant rather than random. **(a)** Image quality (evaluated in terms of the maximum uniformity) as a function of the targeted phase. **(b)** Results with $C_{m,n}^{SIMO}$ in terms of normalized amplitude and phase of the obtained holographic image. **(c)** The corresponding results by the GS algorithm with the phase constraint.

$$\widetilde{\phi}_{m,n}(\boldsymbol{r}_q;\boldsymbol{r}_s) = \Delta_{m,n}(\boldsymbol{r}_q;\boldsymbol{r}_s) + \phi(\boldsymbol{r}_q),$$

$$\Delta_{m,n}(\boldsymbol{r}_q;\boldsymbol{r}_s) = k\left(\underbrace{|\boldsymbol{r}_s - \boldsymbol{r}_{m,n}|}_{R_{nm}(r_s)} + \underbrace{|\boldsymbol{r}_q - \boldsymbol{r}_{m,n}|}_{R_{nm}(r_q)}\right).$$

Herein, $\phi_q \equiv \phi(\boldsymbol{r}_q)$ $(0 \leq \phi_q < 2\pi)$ is the intended phase directed from the source to the receiver at $\boldsymbol{r}_q$ through the metasurface. Following similar steps in SIMO, the system response of the one-bit coding metasurface at $\boldsymbol{r}$ can be derived as:

$$\hat{\mathcal{H}}_{MIMO}(\boldsymbol{r},\boldsymbol{r}';\{q_s\}) = \underbrace{E_1^{MIMO}(\boldsymbol{r},\boldsymbol{r}';\{q_s\})}_{\text{leading term}} + \underbrace{\sum\nolimits_{\{p_{qs}\}/\left\{\sum_{\{q_s\}}|p_{q_s}|=1\,\&\,p_{q_s}\neq-1\right\}} E^{MIMO}_{\{p_{qs}\}}(\boldsymbol{r},\boldsymbol{r}';\{q_s\})exp\left[j\sum\nolimits_{s=1}^{S}\sum\nolimits_{q_s=1}^{Q_s} p_{q_s}\phi_{q_s}\right]}_{\text{perturbation terms}} \tag{3.12}$$

The leading term $E_1^{MIMO}(\boldsymbol{r},\boldsymbol{r}';\{q_s\})$ and perturbation term $E^{MIMO}_{\{q_s\}}(\boldsymbol{r},\boldsymbol{r}',\{q_s\})$ are

$$E_1^{MIMO}(\boldsymbol{r},\boldsymbol{r}';\{q_s\}) = B_1^{MIMO}\sum\nolimits_{s=1}^{S}\sum\nolimits_{q_s=1}^{Q_s}\exp\left(j\phi_{q_s}\right)A_1^{SISO}(\boldsymbol{r},\boldsymbol{r}';q,s),$$

and $E^{MIMO}_{\{p_{qs}\}}(\boldsymbol{r},\boldsymbol{r}';\{q_s\}) = B^{MIMO}_{\{p_{qs}\}}A^{MIMO}_{\{p_{qs}\}}(\boldsymbol{r},\boldsymbol{r}';\{q_s\})$, respectively.

Moreover, B_1^{MIMO}, $A^{MIMO}_{\{p_{qs}\}}$, and $B^{MIMO}_{\{p_{qs}\}}$ are defined as

$$B_1^{MIMO} = -\frac{j}{\pi}\int_{-\infty}^{\infty}\frac{1}{\xi}J_0^{\left(\sum_{s=1}^{S}Q_s\right)-1}(\xi)J_1(\xi)d\xi,$$

$$A^{MIMO}_{\{p_{qs}\}}(\boldsymbol{r},\boldsymbol{r}';\{q_s\}) = \sum\nolimits_{m,n}\frac{\exp\left(j\left[\sum_{s=1}^{S}\sum_{q_s=1}^{Q_s}p_{q_s}\Delta_{m,n}(\boldsymbol{r}_q;\boldsymbol{r}_s) - \Delta_{m,n}(\boldsymbol{r};\boldsymbol{r}')\right]\right)}{R_{nm}(\boldsymbol{r})R_{nm}(\boldsymbol{r}')}$$

and $B^{MIMO}_{\{p_{qs}\}} = -\frac{j}{\pi}\int_{-\infty}^{+\infty}\frac{1}{\xi}\prod_{s=1}^{S}\prod_{q_s=1}^{Q_s}j^{p_{qs}}J_{p_{q_s}}(\xi)d\xi$, respectively. Note that $B^{MIMO}_{\{p_{qs}\}} = 0$ when $\sum_{s=1}^{S}\sum_{q_s=1}^{Q_s}|p_{q_s}|$ is an even number. Given the resemblance between **Eq. (3.12) and 3.10**, the conclusions for SIMO are applicable to MIMO.

So far, we have assumed an "ideal" one-bit quantization of the programmable metasurface. Specifically, we assumed that upon switching the meta-atom from its "0" (or "1") state to its "1" (or "0") state, the reflection phase experiences a change of 180° while the reflection amplitude remains unchanged. However, such ideal one-bit quantization cannot be realized in practice for many reasons, including nonideal active lumped elements (e.g., the PIN diode) and fabrication errors. The reflection response of the nonideal one-bit meta-atom is $A_+e^{j\phi_+}$ and $A_-e^{j\phi_-}$ for states "1" and "0" (with $0 \leq A_\pm \leq 1$), respectively, instead of 1 and -1. The methodology developed in our paper can be readily generalized to nonideal one-bit meta-atoms. We discuss the system response of the coding metasurface with nonideal one-bit quantization. Particularly, when the meta-atom is illuminated by a plane wave, its binary response states read:

$$f(x) = \begin{cases} A_+e^{j\phi_+}, & \text{for state}\text{"}1\text{"} \\ A_-e^{j\phi_-}, & \text{for state "}0\text{"} \end{cases}, \tag{3.13}$$

where $0 \leq A_+,\ A_- \leq 1$, and $0 \leq \phi_+, \phi_- < 2\pi$. Recall the step function denoted by $u(x) = \begin{cases} 1, & x > 0 \\ \frac{1}{2}, & x = 0 \\ 0, & x < 0 \end{cases}$, then the above piece-wise function is expressed as

$$f(x) = A_+e^{j\phi_+}u(x) + A_-e^{j\phi_-}u(-x). \tag{3.14}$$

Immediately, **Eq. (3.14)** can be further expressed as

$$f(x) = \underbrace{\frac{\left(A_+e^{j\phi_+} + A_-e^{j\phi_-}\right)}{2}}_{\gamma_0} \underbrace{\frac{\left(A_+e^{j\phi_+} - A_-e^{j\phi_-}\right)}{2}}_{\gamma_1} \text{sign}(x). \tag{3.15}$$

It is clear that we have established a simple relation of the responses between the ideal one-bit meta-atom and the nonideal meta-atom.

For the nonideal one-bit coding metasurface, its control coding patterns for SISO, SIMO, and MIMO can be derived as

$$C_{m,n}^{non\text{-}ideal,SISO} = \gamma_0 + \gamma_1 C_{m,n}^{SISO}, \tag{3.16a}$$

$$C_{m,n}^{non\text{-}ideal,SIMO} = \gamma_0 + \gamma_1 C_{m,n}^{SIMO}, \tag{3.16b}$$

$$C_{m,n}^{non\text{-}ideal,MIMO} = \gamma_0 + \gamma_1 C_{m,n}^{MIMO}. \tag{3.16c}$$

For SISO, SIMO, and MIMO, the system responses of the coding metasurface with the nonideal one-bit quantization can be readily derived as

$$\hat{\mathcal{H}}_{SISO}^{non\text{-}ideal}(\mathbf{r},\mathbf{r}';q,s) = \gamma_0 \sum\nolimits_{m,n} \frac{\exp[-jk\Delta_{nm}(\boldsymbol{r};\mathbf{r}')]}{R_{nm}(\mathbf{r})R_{nm}(\mathbf{r}')} + \gamma_1 \hat{\mathcal{H}}_{\text{SISO}}(\mathbf{r},\mathbf{r}';q,s),$$

$$\hat{\mathcal{H}}_{SIMO}^{non\text{-}ideal}(\mathbf{r},\mathbf{r}';\{q\},s) = \gamma_0 \sum\nolimits_{m,n} \frac{\exp[-jk\Delta_{nm}(\boldsymbol{r};\mathbf{r}')]}{R_{nm}(\boldsymbol{r})R_{nm}(\mathbf{r}')} + \gamma_1 \hat{\mathcal{H}}_{\text{SIMO}}(\mathbf{r},\mathbf{r}';\{q\},s),$$

$$\hat{\mathcal{H}}_{MIMO}^{non\text{-}ideal}(\mathbf{r},\mathbf{r}';\{q_s\}) = \gamma_0 \sum\nolimits_{m,n} \frac{\exp[-jk\Delta_{nm}(\boldsymbol{r};\mathbf{r}')]}{R_{nm}(\boldsymbol{r})R_{nm}(\mathbf{r}')} + \gamma_1 \hat{\mathcal{H}}_{\text{MIMO}} \hat{\mathcal{H}}_{\text{MIMO}}(\mathbf{r},\mathbf{r}';\{q_s\}).$$

3.2 Information Capacities of One-Bit Coding Metasurfaces

In this section, we investigate the relation between the information capacities of a metasurface with one-bit coding and a metasurface with continuous coding. The one-bit quantization of the coding metasurface gives rise to the quantization-related signal interferences between different channels in SIMO and MIMO settings. In the context of wireless communication, this yields so-called channel cross talk; in the context of sensing, this results in image blurring. We now study this effect for concreteness in SIMO, but the developed results can easily be extended to MIMO. Throughout this section, we assume that the source and receivers are not in the "blind district," that is, the perturbation terms are negligible compared to the leading term. We consider an information stream with M_i-level PSK modulation, where the source emits waves and the PSK information modulation is accomplished upon interaction with the metasurface, *and* the waves are directed to the desired receivers. We stress that it is not the source but the metasurface that performs the PSK modulation. The list of possible desired phases at the ith receiver (located at $\boldsymbol{r}_i$) is

$$\phi_i \in \theta_i + \left\{0, \frac{2\pi}{M_i}, 2\frac{2\pi}{M_i}, \ldots, (M_i - 1)\frac{2\pi}{M_i}\right\}, i = 1, 2, \ldots, Q,$$

where the symbol θ_i denotes the phase bias for the ith receiver.

We here consider a realistic scenario, where any two receivers are well separated in terms of the Rayleigh limit, $|A_1^{SISO}(\boldsymbol{r}_i, \boldsymbol{r}_s; q, s)| \approx 0$ for $i \neq q$. As detailed in **Appendix 3**, we can obtain the coherence of $\hat{\mathcal{H}}_{SIMO}(\boldsymbol{r}_i, \boldsymbol{r}_s; \{q\}, s)$ and $\hat{\mathcal{H}}_{SIMO}\left(\boldsymbol{r}_j, \boldsymbol{r}_s; \{q\}, s\right)$:

$$\hat{\mathcal{H}}_{SIMO}(\boldsymbol{r}_i, \boldsymbol{r}_s;\ \{q\}, s), \hat{\mathcal{H}}_{SIMO}(\boldsymbol{r}_j, \boldsymbol{r}_s;\ \{q\}, s) = B_0^{SIMO} A_0 \sum\nolimits_{n_{p:1\to Q}/n_i} \mathcal{Z}^*_{\{M_p n_p \theta_p\}} E^{SIMO,*}_{\{M_p n_p\}}(\boldsymbol{r}_j; \{q\}, s)$$

$$+B_0^{SIMO,*}A_0^*\sum\nolimits_{n_{p:1\to Q}/n_i}\mathcal{Z}_{\{M_p n_p \theta_p\}}E_{\{M_p n_p\}}^{SIMO}(\boldsymbol{r}_i;\{q\},s) \quad \text{for } i\neq j \tag{3.17}$$

and $|\hat{\mathcal{H}}_{SIMO}(\boldsymbol{r}_i,\boldsymbol{r}_s;\{q\},s)|^2\approx|B_0^{SIMO}A_0|^2$.

Herein, $\mathcal{Z}_{\{p_q\theta_q\}}=\exp\left(j\sum_q p_q\theta_q\right)$, and we have assumed that $A_0\approx A_1^{SISO}(\boldsymbol{r}_i,\boldsymbol{r}_s;i,s)$ holds for any $\boldsymbol{r}_i\in\{q\}$ when the receivers are in a relative small range. We can observe that the coherence function depends on the choice of the phase bias $\{\theta_i\}$, implying that the quantization-related signal interference can be improved by optimizing the setting of the phase bias $\{\theta_i\}$. This observation is consistent with the results presented at the end of this section in the context of holography.

To rigorously evaluate additional signal interference originating from the one-bit quantization of the metasurface, we now examine the statistical behavior of a given digital information symbol transferred through a given wireless SIMO channel while simultaneously the remaining SIMO links are deployed for other (statistically independent) digital information streams. For the digital information symbol acquired at $\boldsymbol{r}_u$, the statistical mean $\mu_u^{SIMO}=\hat{\mathcal{H}}_{SIMO}(\boldsymbol{r}_u,\boldsymbol{r}_s;\{q\},s)$ and variance $\sigma_u^{SIMO}=\langle|\hat{\mathcal{H}}_{SIMO}(\boldsymbol{r}_u,\boldsymbol{r}_s;\{q\},s)-\langle\hat{\mathcal{H}}_{SIMO}(\boldsymbol{r}_u,\boldsymbol{r}_s;\{q\},s)\rangle|^2\rangle$ are readily derived as

$$\begin{aligned}\mu_u^{SIMO}&=B_0^{SIMO}\underbrace{A_1^{SISO}(\boldsymbol{r}_u,\boldsymbol{r}_s;u,s)\exp(j\phi_u)}_{\textit{for continuous coding metasurface}}+\\&\underbrace{\sum\nolimits_{\{p_q\}/\{\sum_q|p_q|=1\&p_q\#-1\}}\hat{\mathcal{H}}_{\{p_q\}}^{SIMO}(\boldsymbol{r}_u,\boldsymbol{r}_s;\{q\},s)exp[jp_u\phi_u]\mathcal{Z}_{\{p_q\theta_q\}/p_u\theta_u}\prod\nolimits_{q=1,q\#u}^{Q}\delta_{p_q-n_q M_q}}_{\textit{one-bit quantization mean bias}}\end{aligned} \tag{3.18a}$$

and

$$\sigma_{SIMO}^2|B_0^{SIMO}|^2\underbrace{\sum\nolimits_{q,q\neq u}\sigma_q^2A_1^{SISO}(\boldsymbol{r}_u,\boldsymbol{r}_s;q,s)^2}_{\textit{for continuous coding metasurface}}+\underbrace{\chi(u,\{M_q\})}_{\textit{one-bit quantization variance}}. \tag{3.18b}$$

Herein, $\mathcal{Z}_{\{p_q\theta_q\}/p_u\theta_u}=\exp\left(j\sum_{q,q\neq u}p_q\theta_q\right)$, σ_q^2 is the covariance of the intended information stream of the qth receiver. In addition, the factor $\chi(u,\{M_q\})$ characterizes the signal interferences arising from the one-bit quantization of the metasurface. Furthermore, for the whole SIMO system, the variance can be approximated by taking the average of $\{\sigma_u^{SIMO}\}$ over all intended receivers. As a result, we have

$$\sigma_{SIMO}^2 = \frac{1}{Q}\sum_{u=1}^{Q}\sigma_u^{SIMO}$$
$$= |B_0^{SIMO}|^2 \underbrace{\frac{1}{Q}\sum_{u=1}^{Q}\sum_{q=1,q\neq u}^{Q}\sigma_q^2 A_1^{SISO}(\boldsymbol{r}_u, \boldsymbol{r}_s; q, s)^2}_{\sigma_{cont}^2}$$
$$+ \underbrace{\frac{1}{Q}\sum_{u=1}^{Q}\chi(u, \{M_q\})}_{\chi}. \quad (3.19)$$

Now, it is clear that if an continuous coding metasurface is used instead, variance and mean are characterized by σ_{cont}^2 and $A_1^{SISO}(\boldsymbol{r}_u; u, s)\exp(j\phi_u)$. Therefore, the effect of the one-bit quantization of the metasurface is obvious from **Eq. (3.19)**: the one-bit quantization-induced mean bias as well as the additional noise due to the one-bit quantization variance both reduce the distinguishability of distinct information symbols.

Based on Shannon's theory, the information capacity C_{cont} of a continuous coding metasurface can be readily obtained [51–52]. For instance, for the SIMO case, it reads $C_{cont} = Q\log_2\left(1 + S/\left(\sigma_n^2 + \sigma_{cont}^2\right)\right)$ per frequency, where S and σ_n^2 denote the signal and system noise levels, respectively, and $S = \mathcal{P}|A_1^{SISO}(\boldsymbol{r}_q, \boldsymbol{r}_s; q, s)|^2 \approx \mathcal{P}|MN/r_s r_q|^2$ (far-field approximation). The one-bit quantization of the metasurface has two consequences: (i) an energy loss by a factor of $\gamma_{one-bit}^2$, and (ii) additional noise $\sigma_{one-bit}^2$. Both effects deteriorate the SNR and thereby reduce the information capacity in the case of the one-bit coding metasurface:

$$C_{one-bit} \approx Q\log_2\left(1 + \gamma_{one-bit}^2 S/\left(\sigma_n^2 + \sigma_{one\text{-}bit}^2\right)\right) \quad (3.20)$$

where $\gamma_{one\text{-}bit} = \begin{cases} B_1^{SISO}, & for\ SISO \\ B_1^{SIMO}, & for\ SIMO \\ B_1^{MIMO}, & for\ MIMO \end{cases}$,

and $\sigma_{one\text{-}bit}^2 = \begin{cases} 0, & for\ SISO \\ \sigma_{SIMO}^2, & for\ SIMO \\ \sigma_{MIMO}^2, & for\ MIMO \end{cases}$.

Now, taking the SIMO as an illustrative example, we can derive the relation between C_{cont} and $C_{one\text{-}bit}$ as

$$\frac{\Delta C}{Q} = \frac{C_{one\text{-}bit} - C_{cont}}{Q} \approx \log_2\left(\gamma_{one\text{-}bit}^2 \frac{\sigma_n^2 + \sigma_{cont}^2}{\sigma_n^2 + \sigma_{one\text{-}bit}^2}\right)$$
$$\approx \log_2\left(\gamma_{one\text{-}bit}^2 \frac{\sigma_n^2 + \sigma_{cont}^2}{\sigma_n^2 + \gamma_{one\text{-}bit}^2\sigma_{cont}^2 + \chi}\right) \quad (3.21)$$

where we assumed $SNR \gg 1$ and $\sigma^2_{one\text{-}bit} = \gamma^2_{one\text{-}bit}\sigma^2_{cont} + \chi$ in **Eq. (3.21)**. Considering that $\frac{\Delta C}{Q}$ behaves as a monotone decreasing function of σ_n^2 for $\sigma^2_{cont} > \sigma^2_{one\text{-}bit}$, we can estimate its lower and upper bounds:

$$\log_2\left(\gamma^2_{one\text{-}bit}\right) \leq \frac{\Delta C}{Q} \leq \log_2\left(\frac{\gamma^2_{one\text{-}bit}\sigma^2_{cont}}{\gamma^2_{one\text{-}bit}\sigma^2_{cont} + \chi}\right) \tag{3.22}$$

The upper bound of $\frac{\Delta C}{Q}$ is reached when the system noise level σ_n^2 is relatively low ($\sigma_n^2 \ll \sigma^2_{cont}$ and $\sigma_n^2 \ll \gamma^2_{one\text{-}bit}\sigma^2_{cont} + \chi$). Furthermore, if $\chi \ll \gamma^2_{one\text{-}bit}\sigma^2_{cont}$, the upper bound approaches zero, that is, $\frac{\Delta C}{Q} \to 0$, which implies that the one-bit coding metasurface has nearly the same information capacity as the continuous metasurface under these conditions.

We go on to evaluate the normalized difference in information capacities $\frac{\Delta C}{Q}$ of the one-bit and continuous coding metasurfaces in the SIMO scenario. **Figure 3.6a** plots $\frac{\Delta C}{Q}$ as a function of the number of channels Q for different system SNRs, and **Fig. 3.6b** compares $\sigma^2_{one\text{-}bit}$ and σ^2_{cont} as a function of the number of channels Q, in which the source and the receivers are deployed in the near-field region of the metasurface. It can be verified from this set of figures that $\frac{\Delta C}{Q}$ approaches the upper bound identified in **Eq. (3.22)**, and that the upper bound will tend to zero, as the system SNR is increased; in contrast, for low system SNRs the lower bound $\log_2\left(\gamma^2_{one\text{-}bit}\right)$ is approached. Some representative constellation diagrams for $Q = 3$ and $Q = 6$ for different system SNRs are shown in **Fig. 3.6c and d**, in which the red and black stars characterize the signal interferences for the system-noise-free case, respectively. Additionally, the bit-error analysis on the system of one-bit coding metasurface can be made along the same line. It is obvious, in line with our previous calculations, that the one-bit quantization of the metasurface will give rise to additional signal interferences, and that such interferences will become stronger if the number of channels Q is larger. Nonetheless, it can be observed that the one-bit coding programmable metasurface performs almost as well as the continuous coding metasurface as long as Q is not too large.

Finally, we briefly evaluate the performance of the one-bit coding metasurface in terms of the bit error rate (BER) in the SISO scenario for different phase quantization levels M, in which the source and receiver are deployed in the near field of metasurface. **Figure 3.7a** shows the dependence of BER on M with the system SNR of 20dB. **Figure 3.7b and c** reports the representative constellation diagrams for $M = 3$ and $M = 4$ with different system channel numbers, respectively. For comparison, the corresponding results of the continuous coding metasurface are also provided. Overall, these results illustrate that the one-bit coding metasurface's BER performance can be comparable to that of the continuous metasurface in PSK with low values of M.

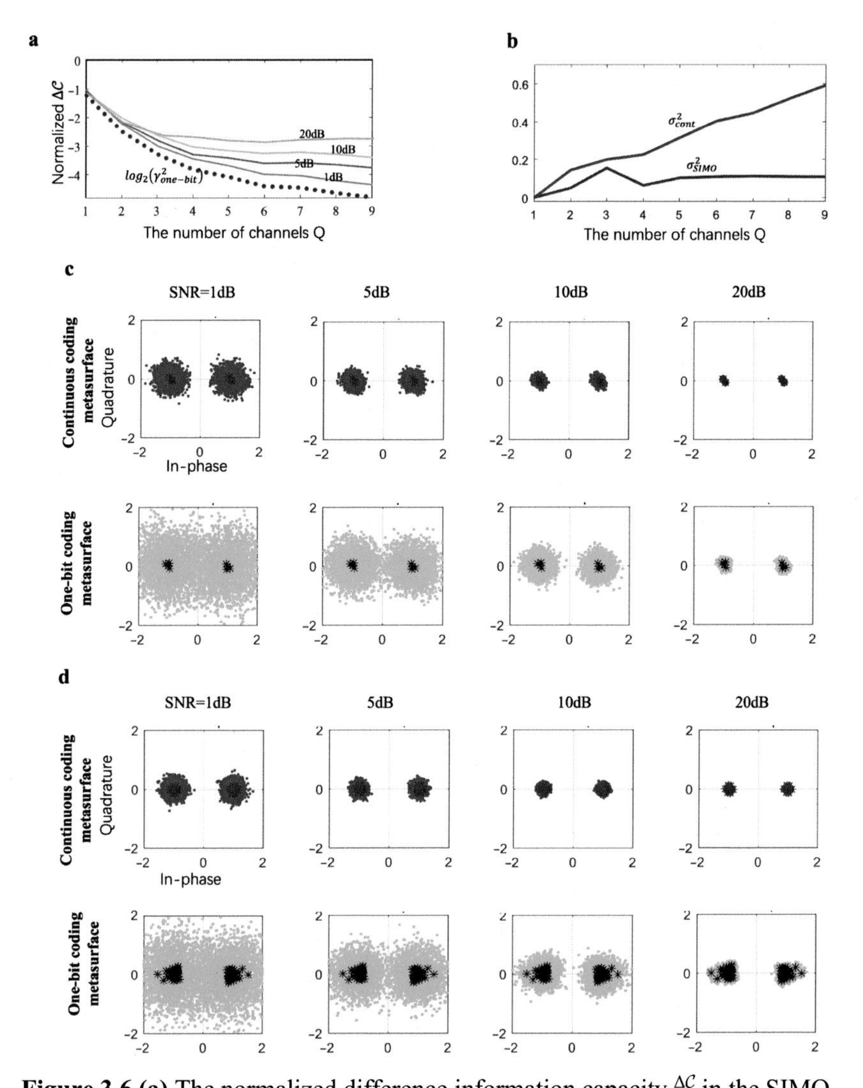

Figure 3.6 (a) The normalized difference information capacity $\frac{\Delta C}{Q}$ in the SIMO setting, in comparing the one-bit coding metasurface with the continuous metasurface in the SIMO scenario as a function of Q for different system SNRs when the receiver and source are in the near-field region of the metasurface. For comparison, the lower bound of $\frac{\Delta C}{Q}$ represented by $\log_2\left(\gamma^2_{one\text{-}bit}\right)$ is also plotted. The binary PSK ($M = 2$) is considered. **(b)** The comparison of $\sigma^2_{one\text{-}bit}$ and σ^2_{cont} as a function of the number of channels Q. **(c–d).** Constellation diagrams for $Q = 3$ and $Q = 6$ for different system SNRs. The constellation diagram axes are normalized by B_0^{SIMO}. Additionally, the red-marked and black-marked stars of the constellation diagrams characterize the signal interferences for the system-noise-free case, respectively. In this set of examinations, the source is located at (0,0,1.2 m), and the receivers are deployed at the distance of 3 m away from the metasurface.

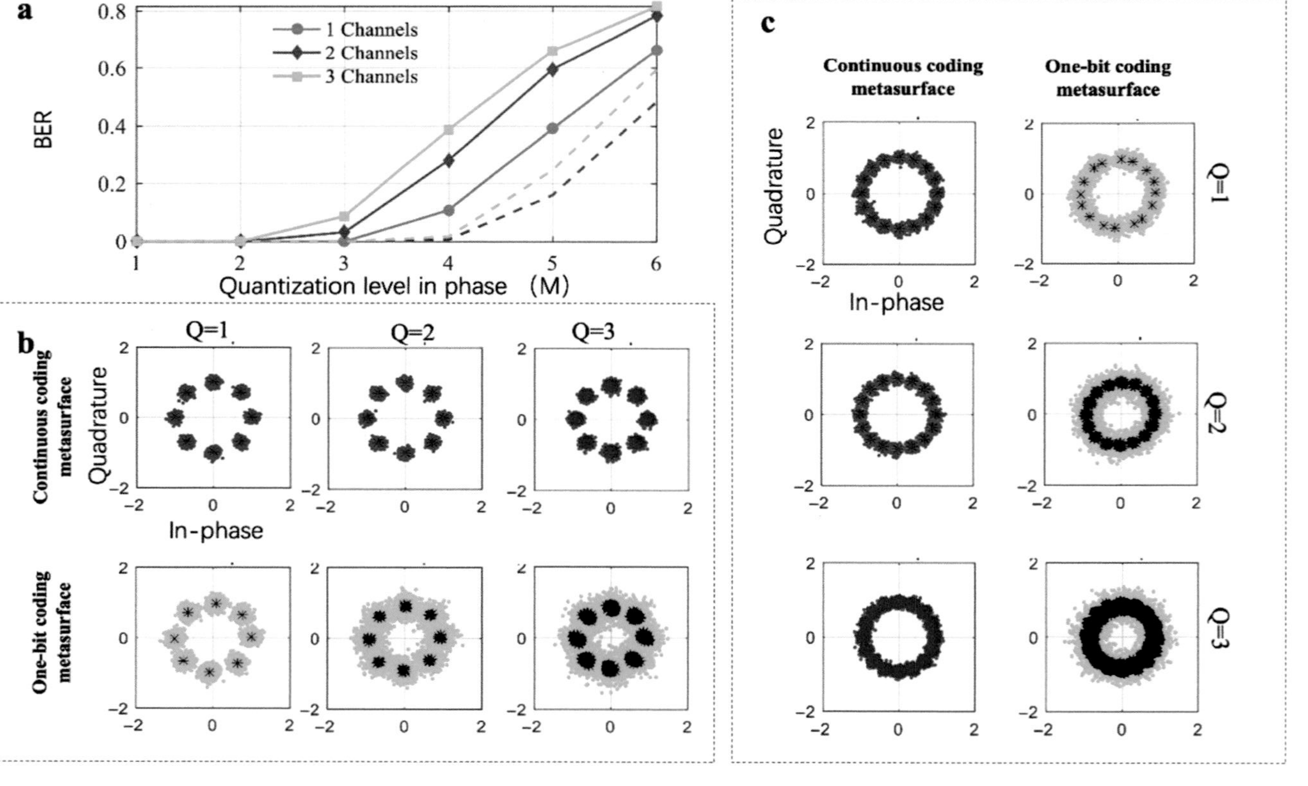

Figure 3.7 Bit error rate (BER) comparison of the one-bit and continuous coding metasurface in the SISO setting. **(a)** Dependence of BER on the phase quantization level M in a system with an SNR of 20 dB, in which the receiver and the source are not in the "blind district" of the metasurface, in which the solid and dashed lines represent the cases of continuous and one-bit coding metasurface. **(b,c)** Constellation diagrams for different phase quantization levels, $M = 3$ **(b)** and $M = 4$ **(c)**, for different Channel Number $Q = 1, 2, 3$ in the cases of one-bit and continuous coding, in which the axes are normalized by B_0^{SIMO}. The red-marked and black-marked stars of the constellation diagrams in **(b)** and **(c)** correspond to the system-noise-free case.

To summarize, we have explored fundamental limitations of the one-bit coding metasurfaces in comparison to the continuous coding metasurfaces in terms of their information capacity and considered illustrative examples from key applications in wireless communications and holography. Our results illustrate surprisingly nearly no performance deterioration due to the one-bit coding under mildly favorable constraints, such as low-level PSK, few-user SIMO, or relaxed phase constraints on holograms. We expect that these fundamental insights will impact a wide range of metasurface-assisted techniques seeking to control the flow of information, both for electromagnetic waves and other frequencies and wave phenomena.

Appendix 1 Derivation of Eq. (3.2)

We here elaborate on the derivation of **Eq. (3.2)** and define the involved notations. For the SISO setting, the control coding pattern of the one-bit coding metasurface reads

$$C_{m,n}^{SISO} = \mathrm{sign}\left[\cos\left(\tilde{\phi}_{nm}^{SISO}\right)\right], \tag{A1}$$

where $\tilde{\phi}_{nm}^{SISO} \equiv \tilde{\phi}_{m,n}(\boldsymbol{r}_q; \boldsymbol{r}_s)$

$$\tilde{\phi}_{m,n}(\boldsymbol{r}_q; \boldsymbol{r}_s) = \Delta_{m,n}(\boldsymbol{r}_q; \boldsymbol{r}_s) + \phi(\boldsymbol{r}_q),$$

$$\Delta_{m,n}(\boldsymbol{r}_q; \boldsymbol{r}_s) = k\left(\underbrace{|\boldsymbol{r}_s - \boldsymbol{r}_{m,n}|}_{R_{nm}(r_s)} + \underbrace{|\boldsymbol{r}_q - \boldsymbol{r}_{m,n}|}_{R_{nm}(r_q)}\right).$$

Herein, $\phi_q \equiv \phi(\boldsymbol{r}_q)$ $(0 \leq \phi_q < 2\pi)$ denotes the intended phase directed from the source to the receiver at $\boldsymbol{r}_q$ through the metasurface, and $\mathrm{sign}(x)$ is the sign function:

$$\mathrm{sign}(x) = \begin{cases} +1, & x > 0 \\ 0, & x = 0. \\ -1, & x < 0 \end{cases}$$

Using the following identical equations:

$$\mathrm{sign}(x) = \frac{-j}{\pi}\int_{-\infty}^{\infty}\frac{\exp(jx\xi)}{\xi}d\xi$$

and

$$\exp\left(j\xi\cos\left(\widetilde{\phi}_{nm}\right)\right) = \sum_{p=-\infty}^{\infty}\exp\left(jp\widetilde{\phi}_{nm}\right)j^{p}J_{p}(\xi),$$

we can express **Eq. (A1)** as:

$$\begin{aligned} C_{m,n}^{SISO} &= \frac{-j}{\pi}\int_{-\infty}^{\infty}\frac{1}{\xi}\sum_{p=-\infty}^{\infty}\exp\left(jp\widetilde{\phi}_{nm}\right)j^{p}J_{p}(\xi)d\xi \\ &= \sum_{p=-\infty}^{\infty}B_{1}^{SISO}\exp\left(jp\widetilde{\phi}_{nm}\right), \end{aligned} \tag{A2}$$

where

$$\begin{aligned} B_{1}^{SISO} &= \frac{-j^{p+1}}{\pi}\int_{-\infty}^{\infty}\frac{1}{\xi}J_{p}(\xi)d\xi \\ &= \frac{-j^{p+1}}{\pi}\int_{-\infty}^{\infty}\frac{1}{\xi}J_{p}(\xi)d\xi \\ &= \begin{cases} \frac{-j^{p+1}}{\pi}\cdot\frac{2}{p}, & p \text{ is odd} \\ 0, & \text{else} \end{cases}. \end{aligned}$$

Now we proceed to derive the radiation response of the one-bit coding metasurface. To that end, the so-called discrete dipole approximation is applied,[59] and each macro meta-atom is modeled as an ideal isotropic dipole. As pointed out in Section 2, the macro meta-atoms are modeled as having mutually independent EM responses. Then, the radiation response of metasurface $\hat{\mathcal{H}}_{\mathrm{SISO}}(\mathbf{r},\mathbf{r}';q,s)$ reads

$$\hat{\mathcal{H}}_{\mathrm{SISO}}(\mathbf{r},\mathbf{r}';q,s) = \sum_{m,n}C_{m,n}^{SISO}\cdot\frac{\exp\left(-jkR_{nm}(\mathbf{r})\right)}{R_{nm}(\mathbf{r})}\frac{\exp[-jkR_{nm}(\mathbf{r}'))]}{R_{nm}(\mathbf{r}')},$$

which holds up to a multiplicative factor. It is noted that the scalar approximation has been explicitly assumed for simplicity; however, more general full vectorial discussions could be made following the presented strategy along with the use of the dyadic Green's function [58,59]. Then, using Eq. (A1) in the preceding equation leads to

$$\hat{\mathcal{H}}_{\text{SISO}}(\mathbf{r},\mathbf{r}';q,s) = \sum\nolimits_{m,n} C_{m,n}^{SISO} \cdot \frac{\exp[-j\Delta_{nm}(\boldsymbol{r};\mathbf{r}'))]}{R_{nm}(\boldsymbol{r})R_{nm}(\mathbf{r}')}$$

$$= B_1^{SISO}\exp\left(j\phi_q\right)\underbrace{\sum\nolimits_{m,n}\frac{\exp[j\Delta_{m,n}(\boldsymbol{r}_q;\boldsymbol{r}_s) - j\Delta_{nm}(\boldsymbol{r};\boldsymbol{r}'))]}{R_{nm}(\boldsymbol{r})R_{nm}(\mathbf{r}')}}_{A_1^{SISO}(\boldsymbol{r},\mathbf{r}';q,s)}$$

$$+\sum\nolimits_{p=-\infty,\ p\neq 1}^{\infty} B_p^{SISO}\exp\left(jp\phi_q\right)\underbrace{\sum\nolimits_{m,n}\frac{\exp[jp\Delta_{m,n}(\boldsymbol{r}_q;\boldsymbol{r}_s) - j\Delta_{nm}(\boldsymbol{r};\boldsymbol{r}'))]}{R_{nm}(\boldsymbol{r})R_{nm}(\mathbf{r}')}}_{A_p^{SISO}(\boldsymbol{r},\boldsymbol{r}';q,s)} \tag{A3}$$

where $\sum_{m,n} \equiv \sum_{n=1}^{n}\sum_{m=1}^{M}$.

After introducing the following notations,

$$E_p^{SISO}(\boldsymbol{r},\mathbf{r}';q,s) = B_p^{SISO}\exp\left(j\phi_q\right)A_p^{SISO}(\boldsymbol{r},\mathbf{r}';q,s)$$

$$A_p^{SISO}(\boldsymbol{r},\mathbf{r}';q,s) = \sum\nolimits_{m,n}\frac{\exp[jp\Delta_{m,n}(\boldsymbol{r}_q;\mathbf{r}_s) - j\Delta_{nm}(\boldsymbol{r};\mathbf{r}'))]}{R_{nm}(\boldsymbol{r})R_{nm}(\mathbf{r}')},$$

the proof of Eq. (3.2) can be readily completed.

Under the far-field assumptions [58], we have

$$\Delta_{m,n}(\boldsymbol{r}_q;\boldsymbol{r}_s) = k(|\boldsymbol{r}_s - \boldsymbol{r}_{m,n}| + |\boldsymbol{r}_q - \boldsymbol{r}_{m,n}|)$$

$$\approx k\left(r_s - \hat{\boldsymbol{r}}_s\cdot\boldsymbol{r}_{m,n} + r_q - \hat{\boldsymbol{r}}_q\cdot\boldsymbol{r}_{m,n}\right),$$

$$\Delta_{m,n}(\boldsymbol{r};\boldsymbol{r}_s) = k(|\boldsymbol{r} - \boldsymbol{r}_{m,n}| + |\boldsymbol{r}_s - \boldsymbol{r}_{m,n}|)$$

$$\approx k\left(r - \hat{\mathbf{r}}\cdot\boldsymbol{r}_{m,n} + r_s - \hat{\boldsymbol{r}}_s\cdot\boldsymbol{r}_{m,n}\right),$$

$\frac{1}{R_{nm}(\boldsymbol{r}')} \approx \frac{1}{r'}$, and $\frac{1}{R_{nm}(\boldsymbol{r})} \approx \frac{1}{r}$.

Then, the term $A_p^{SISO}(\boldsymbol{r},\boldsymbol{r}_s;q,s)$ can be expressed as

$$A_p^{SISO}(\boldsymbol{r},\boldsymbol{r}_s;q,s) = \sum\nolimits_{m,n}\frac{\exp\left[j\left(p\Delta_{nm}(\boldsymbol{r}_q;\boldsymbol{r}_s) - \Delta_{nm}(\boldsymbol{r};\boldsymbol{r}_s)\right)\right]}{R_{nm}(\boldsymbol{r})R_{nm}(\boldsymbol{r}_s)}$$

$$\approx \frac{\exp\left[jk\left(pr_q - r + (p-1)r_s\right)\right)\right]}{rr_s}\sum\nolimits_{m,n}\exp\left[jk\left((1-p)\hat{\boldsymbol{r}}_s - p\hat{\boldsymbol{r}}_q + \hat{\boldsymbol{r}}\right)\Delta\boldsymbol{r}_{m,n})\right]$$

Specifically, if $\hat{\boldsymbol{r}} - p\hat{\boldsymbol{r}}_q + (1-p)\hat{\boldsymbol{r}}_s \parallel \hat{\boldsymbol{n}}$, then $A_p^{SISO}(\boldsymbol{r},\boldsymbol{r}';q,s)$ becomes

$$A_p^{SISO}(\boldsymbol{r}, \boldsymbol{r}_s; q, s) \approx MN \frac{\exp\left[jk\left(pr_q - r + (p-1)r_s\right)\right]}{rr_s}.$$

Appendix 2 Derivation of Eq. (3.10)

For the SIMO setting, the coding pattern of the one-bit coding metasurface reads:

$$C_{m,n}^{SIMO} = \text{sign}\left[\sum_{q=1}^{Q} \cos\left(\widetilde{\phi}_{nm}^{SIMO}(q)\right)\right], \quad \text{(A4)}$$

where $\widetilde{\phi}_{nm}^{SIMO}(q) \equiv \widetilde{\phi}_{m,n}\left(\boldsymbol{r}_q; \boldsymbol{r}_s\right)$,

$$\widetilde{\phi}_{m,n}\left(\boldsymbol{r}_q; \boldsymbol{r}_s\right) = \Delta_{m,n}\left(\boldsymbol{r}_q; \boldsymbol{r}_s\right) + \phi_q,$$

and $\Delta_{m,n}\left(\boldsymbol{r}_q; \boldsymbol{r}_s\right) = k\left(\underbrace{|\boldsymbol{r}_s - \boldsymbol{r}_{m,n}|}_{R_{nm}(\boldsymbol{r}_s)} + \underbrace{|\boldsymbol{r}_q - \boldsymbol{r}_{m,n}|}_{R_{nm}(\boldsymbol{r}_q)}\right).$

Similar to **Appendix 1**, we express **Eq. (A4)** as

$$\begin{aligned} C_{m,n}^{SIMO} &= \text{sign}\left[\sum_{q=1}^{Q} \cos\left(\widetilde{\phi}_{nm}^{SIMO}(q)\right)\right] \\ &= \frac{-j}{\pi}\int_{-\infty}^{\infty} \frac{1}{\xi}\prod_{q=1}^{Q} \exp\left[j\xi \cos\left(\widetilde{\phi}_{nm}(q)\right)\right] d\xi \\ &= \sum_{p_1}\sum_{p_2}\cdots\sum_{p_Q} \exp\left(j\sum_{q} p_q \widetilde{\phi}_{nm}(q)\right) \\ &\int_{-\infty}^{\infty} \frac{-j}{\pi} j^{\sum_q p_q} \frac{1}{\xi}\prod_{q=1}^{Q} J_{p\,q}(\xi) d\xi \end{aligned} \quad \text{(A5)}$$

in which, $\sum_{p_i} \equiv \sum_{p_i=-\infty}^{\infty}$ and $\sum_q \equiv \sum_{q=1}^{Q}$.
After introducing the notations

$$B_{\{p_i\}}^{SIMO} \equiv \frac{-j}{\pi} j^{\sum_q p} \int_{-\infty}^{\infty} \frac{1}{\xi}\prod_{q=1}^{Q} J_{p_q}(\xi) d\xi$$

and $\sum_{\{p_i\}} \equiv \sum_{p_1}\sum_{p_2}\cdots\sum_{p_Q}$, we can express **Eq. (A5)** as

$$C_{m,n}^{SIMO} = \sum_{\{p_i\}} B_{\{p_i\}}^{SIMO} \exp\left(j\sum_q p_q \widetilde{\phi}_{nm}(q)\right). \quad \text{(A6)}$$

Now, we can arrive at the SIMO's system response of the one-bit coding metasurface as

$$\hat{\mathcal{H}}_{\text{SIMO}}\left(\mathbf{r},\mathbf{r}';\{q\},s\right)=\sum\nolimits_{m,n}C_{m,n}^{SIMO}\frac{\exp\left[-j\Delta_{nm}\left(\boldsymbol{r};\mathbf{r}'\right)\right)\right]}{R_{nm}\left(\boldsymbol{r}\right)R_{nm}\left(\mathbf{r}'\right)}$$

$$=\sum\nolimits_{\{p_i\}}\exp\left(j\sum\nolimits_{q}p_{q}\phi_{q}\right)B_{\{p_i\}}^{SIMO}\underbrace{\sum\nolimits_{m,n}\frac{\exp\left[j\sum\nolimits_{q}p_{q}\Delta_{m,n}\left(\boldsymbol{r}_{q};\boldsymbol{r}_{s}\right)-j\Delta_{nm}\left(\boldsymbol{r};\mathbf{r}'\right)\right)\right]}{R_{nm}\left(\boldsymbol{r}\right)R_{nm}\left(\mathbf{r}'\right)}}_{A_{\{p_i\}}^{SIMO}(\boldsymbol{r},\mathbf{r}')}$$

$$=B_{1}^{SIMO}\underbrace{\sum\nolimits_{q=1}^{Q}\exp\left(j\phi_{q}\right)A_{1}^{SISO}\left(\boldsymbol{r},\mathbf{r}';q,s\right)}_{E_{1}^{SIMO}\left(\boldsymbol{r},\mathbf{r}';\{q\},s\right)}$$

$$\underbrace{+\sum\{p_i\}/\left\{\sum\nolimits_{i}|p_i|=1\,\&\,p_i\neq-1\right\}E_{\{p_i\}}^{SIMO}\left(\boldsymbol{r},\mathbf{r}';\{q\},s\right)\exp\left[j\sum\nolimits_{q=1}^{Q}p_{q}\phi_{q}\right]}_{perturbation\ terms}\qquad\text{(A7)}$$

Herein, $B_1^{SIMO}=\frac{1}{\pi}\int_{-\infty}^{\infty}\frac{1}{\xi}J_1(\xi)J_0^{Q-1}(\xi)d\xi$,

and $E_{\{p_i\}}^{SIMO}(\boldsymbol{r},\mathbf{r}';\{q\},s)=B_{\{p_i\}}^{SIMO}A_{\{p_i\}}^{SIMO}(\boldsymbol{r},\mathbf{r}')$.

In addition, the operator denoted by $\sum_{\{p_i\}/\left\{\sum_i|p_i|=1\,\&\,p_i\neq-1\right\}}$ is defined as follows. For a function of $f(p_1,p_2,\ldots,p_Q)$, we define

$$\sum\nolimits_{\{p_i\}/\left\{\sum_i|p_i|=1\,\&\,p_i\neq-1\right\}}f(p_1,p_2,\ldots,p_Q)$$
$$=\sum\nolimits_{\{p_i\}}f(p_1,p_2,\ldots,p_Q)-[f(1,0,0,\ldots,0)+f(0,1,0,\ldots,0)+\ldots+f(0,0,0,\ldots,1)].$$

Appendix 3 Derivations of Eqs. (3.17) and (3.18)

We here detail the derivation of **Eq. (3.17)**. In terms of the definition of statistical coherence, we arrive at

$$\langle\hat{\mathcal{H}}_{SIMO}(\boldsymbol{r}_i),\hat{\mathcal{H}}_{SIMO}(\boldsymbol{r}_j)\rangle=|\mathrm{B}_0|^2\sum\nolimits_{q=1}^{Q}\sum\nolimits_{q'=1}^{Q}A_q(\boldsymbol{r}_i)A_{q'}^*(\boldsymbol{r}_j)\left\langle\exp(j(\phi_q-\phi_{q'}))\right\rangle$$
$$+\mathrm{B}_0\sum\nolimits_{q=1}^{Q}A_q(\boldsymbol{r}_i)\sum\nolimits_{p_i:1\to Q}E_{\{p_i\}}^*(\boldsymbol{r}_j)\left\langle\exp\left[j\phi_q-j\sum\nolimits_{q'=1}^{Q}p_{q'}\phi_{q'}\right]\right\rangle$$
$$+B_0^*\sum\nolimits_{q=1}^{Q}A_q^*(\boldsymbol{r}_j)\sum\nolimits_{p_i:1\to Q}E_{\{p_i\}}(\boldsymbol{r}_i)\left\langle\exp\left[j\sum\nolimits_{q'=1}^{Q}p_{q'}\phi_{q'}-j\phi_q\right]\right\rangle$$
$$+\sum\nolimits_{p_i:1\to Q}\sum\nolimits_{p_i':1\to Q}E_{\{p_i\}}(\boldsymbol{r}_i)E_{\{p_i'\}}^*(\boldsymbol{r}_j)\left\langle\exp\left[j\sum\nolimits_{q=1}^{Q}\left(p_q-p_q'\right)\phi_q\right]\right\rangle.\qquad\text{(A8)}$$

Taking the following identical equations into account, namely,

$$\exp\left(j\left(\phi_q - \phi_{q'}\right)\right) = \delta_{q-q'},$$

$$\exp\left[j\sum_{q=1}^{Q} p_{q'}\phi_{q'} - j\phi_q\right] = \delta_{p_q-1-nM_q}\prod_{q'=1,q'\neq q}^{Q}\delta_{p_{q'}-nM_{q'}},$$

$$\exp\left[j\sum_{q=1}^{Q}\left(p_q - p_q'\right)\phi_q\right] = \prod_{q=1}^{Q}\delta_{p_q-p_q'-nM_q},$$

we can rewrite **Eq. (A8)** as

$$\begin{aligned}\hat{\mathcal{H}}_{SIMO}(\boldsymbol{r}_i), \hat{\mathcal{H}}_{SIMO}(\boldsymbol{r}_j) &= |\mathrm{B}_0|^2\sum_{q=1}^{Q} A_q(\boldsymbol{r}_i)A_q^*(\boldsymbol{r}_j)\\ &+ \mathrm{B}_0\sum_{q=1}^{Q} A_q(\boldsymbol{r}_i)\sum_{n_{i:1\to Q}/n_q} E^*_{\{M_i n_i\}}(\boldsymbol{r}_j)\\ &+ B_0^*\sum_{q=1}^{Q} A_q^*(\boldsymbol{r}_j)\sum_{n_{i:1\to Q}/n_q} E_{\{M_i n_i\}}(\boldsymbol{r}_i)\\ &+ \sum_{p_{i:1\to Q}}\sum_{p_i':1\to Q} E_{\{p_i\}}(\boldsymbol{r}_i)E^*_{\{p_i'\}}(\boldsymbol{r}_j)\prod_{q=1}^{Q}\delta_{p_q-p_q'-nM_q}.\end{aligned} \tag{A9}$$

Herein, we have introduced the following notation:

$$\sum_{n_{i:1\to Q}/n_q} \equiv \sum_{n_1=-N,}^{N}\cdots\sum_{n_{q-1}=-N}^{N}\sum_{n_q=-N+1/M_q}^{N+1/M_q}\sum_{n_{q+1}=-N}^{P}\cdots\sum_{n_Q=-N}^{N}.$$

We here consider a more realistic scenario that two receivers are well separated in terms of the Rayleigh limit, namely, $|A_q(\boldsymbol{r}_i)| \approx 0$ for $i \neq q$. Then, **Eq. (A8)** reads

$$\begin{aligned}\hat{\mathcal{H}}_{SIMO}(\boldsymbol{r}_i), \hat{\mathcal{H}}_{SIMO}(\boldsymbol{r}_j) &\approx \mathrm{B}_0 A_i(\boldsymbol{r}_i)\sum_{n_{i:1\to Q}/n_q} E^*_{\{M_i n_i\}}(\boldsymbol{r}_j)\\ &+ B_0^* A_j^*(\boldsymbol{r}_j)\sum_{n_{i:1\to Q}/n_q} E_{\{M_i n_i\}}(\boldsymbol{r}_i), \quad \text{for } i \neq j\\ \hat{\mathcal{H}}_{SIMO}(\boldsymbol{r}_i), \hat{\mathcal{H}}_{SIMO}(\boldsymbol{r}_i) &= |\mathrm{B}_0 A_0|^2 + \mathrm{B}_0 A_i(\boldsymbol{r}_i)\sum_{n_{i:1\to Q}/n_q} E^*_{\{M_i n_i\}}(\boldsymbol{r}_j)\\ &+ B_0^* A_i^*(\boldsymbol{r}_i)\sum_{n_{i:1\to Q}/n_q} E_{\{M_i n_i\}}(\boldsymbol{r}_i).\end{aligned}$$

Now, the derivation of Eq. (3.17) is completed.

Along the same lines as above, we can arrive at the expressions of μ_u^{SIMO} and σ_u^{SIMO}:

$$\begin{aligned}\mu_u^{SIMO} &= \hat{\mathcal{H}}_{SIMO}(\boldsymbol{r}_u, \boldsymbol{r}_s; \{q\}, s)\\ &= \mathrm{A}_0\mathrm{B}_0\exp(j\phi_i) + \sum_{p_{i:1\to Q}}\left(E_{\{p_i\}}(\boldsymbol{r}_i)\exp[jp_i\phi_i]\prod_{q=1,q\neq i}^{Q}\delta_{p_q-nM_q}\right)\end{aligned} \tag{A10}$$

$$\begin{aligned}\sigma_u^{SIMO} &= |\hat{\mathcal{H}}_{SIMO}(\boldsymbol{r}_u, \boldsymbol{r}_s; \{q\}, s) - \hat{\mathcal{H}}_{SIMO}(\boldsymbol{r}_u, \boldsymbol{r}_s; \{q\}, s)|^2\\ &= |\mathrm{B}_0|^2\sum_{q=1,q\neq i}^{Q}\sigma_q^2 A_q(\boldsymbol{r}_i)^2 + \chi(\{M_q\}),\end{aligned} \tag{A11}$$

in which

$$\chi(\{M_q\}) \equiv \sum_{\{p_{i/nM_i}\}} \sum_{\{p_i'/nM_i\}} E_{\{p_i\}}(\boldsymbol{r}_i) E^*_{\{p_i'\}}(\boldsymbol{r}_i) exp[j(p_i - p_i')\phi_i]$$
$$\prod_{q=1}^{Q} \delta_{p_q - p_q' - nM_q}.$$

Now, the derivation of Eq. (3.18) can be completed.

4 Compressive Metasurface Imager

One outstanding limitation of conventional sensing systems is from the sustainable throughput of the imager's memory, exposure time, and illumination conditions. In this section, we focus on the compressive imagers with metasurfaces, called compressive metasurface imagers (CMIs). As their name implies, the CMI has the same operational mechanism as the famous compressive imager, which relies on three critical factors [53–57]: (i) the scene under investigation has a low-dimensional representation, allowing for the sparse representation itself or in a transformed domain; (ii) a single or a few receivers along with a randomly coded mask (i.e., an information-coded metasurface here) are deployed to acquire the information of a probed scene in a compressive manner; (iii) the sparsity-aware reconstruction algorithm is utilized to retrieve the information of a probed scene from the highly reduced measurements. One critical issue of compressive imagers is the construction of cost-efficient controllable masks for wavefront shaping and thus the information compression on the physical level, which can be realized efficiently with the reprogrammable coding metasurface. In this section, we discuss two recent advances made in this area, namely the single-frequency multiple-shot compressive imager, and the wide-band single-shot compressive imager. We expect that these results can generate important impacts in model-based EM sensing and beyond.

4.1 Introduction to Compressive Imagers

Over the past decade, compressive imagers or coded-aperture imagers in combination with sparsity-regularized reconstruction algorithms have gained intensive attention. Typically, a compressive imager relies on the use of a sequence of random masks, by which the modulated information of the probed object is fully captured by a single fixed sensor. When the probed object allows for a low-dimensional representation in a certain transformed domain, either prespecified or trained such as DCT or wavelet, such a single-sensor imager benefits from the fundamental fact that the number of measurements could be drastically reduced compared to that required by conventional imaging techniques. Then, the information of the probed object can be retrieved from a remarkably reduced number of measurements by solving a tractable sparsity-aware optimization problem. In other words, the required measurements (corresponding to the number of the random masks) could be significantly fewer than the unknowns to be reconstructed, which is theoretically guaranteed by the well-known compressed sensing (CS) theory [58–67].

In the spirit of CS, many compressive imagers have been developed. The pioneering work is the well-known optical single-pixel camera invented at Rice University [53]. Later, many interesting extensions and variants on compressive imagers have been proposed. For instance, Hunt et al. [57, 68] and Li et al. [69–72] invented the single-pixel microwave imager, demonstrating its promising potential in generating high-frame-rate and high-quality imaging. Central to compressive imagers, two fundamental questions are: (1) How does the structure information of the imaging scene affect imaging performance? (2) What is the ultimate limit on the imaging performance if the structure information in the imaging procedure is explored? To date, a body of theories has been established. More specifically, various theories have been established to study the performance of model-based signal representation and recovery or, more strictly, the sparse signal recovery, for example, the coherence-based analysis, null-space-based analysis, and RIP (Restricted Isometry Property)- analysis. Nonetheless, we adopt the RIP-based theory for model-based signal recovery, since its bound on the estimation error is tightest over others in general. The RIP implies incoherence and that the linear measurement operator $\boldsymbol{A}$ approximately perserves the geometry of all sparse vectors. Formally, it defines the restricted isometry constant δ_k to be the smallest nonnegative constant such that [58]

$$(1-\delta_k)||\boldsymbol{x}||_2^2 \leq ||\boldsymbol{A}\boldsymbol{x}||_2^2 \leq (1+\delta_k)||\boldsymbol{x}||_2^2 \text{ for all } k \text{ sparse vectors } x \tag{4.1}$$

The Candes–Romberg–Tao theorem is stated as follows:

THEOREM 1 THE CANDES–ROMBERG–TAO THEOREM ([58])

Let $\boldsymbol{A}$ be a measurement operator that satisfies the RIP. Then, for any signal $\boldsymbol{x}$ and its noisy measurement $\boldsymbol{y}=\boldsymbol{A}\boldsymbol{x}+\boldsymbol{n}$ with $||\boldsymbol{n}||_2 \leq \varepsilon$, the solution $\hat{\boldsymbol{x}}$ to Eq. (4.2) satisfies

$$||\hat{\boldsymbol{x}}-\boldsymbol{x}||_2 \leq C\left[\varepsilon + \frac{||\boldsymbol{x}_k-\boldsymbol{x}||_2}{\sqrt{k}}\right],$$

where $\boldsymbol{x}_k$ denotes the vector of the k largest coefficients in magnitude of $\mathbf{x}$. Equation (4.2), used in the preceding calculation, is defined as follows:

$$min_x||\boldsymbol{x}||_1, \qquad \text{s.t.}, ||\boldsymbol{y}-\boldsymbol{A}\boldsymbol{x}||_2 \leq \varepsilon. \tag{4.2}$$

The compressive sensing by itself has arrived at a relatively mature level with a solid body of theories and algorithms. However, its interactions with EM imaging remain challenging, and many important issues deserve to be studied in-depth:

First, the development of the next-generation imaging system with low-cost, low-complexity and high efficiency requires optimal design of the waveform, the programmable or reconfigurable antenna, and the configuration of sparse sensor array.

Second, it is desirable to establish easily implemented imaging formulations, which account for the real interaction between the EM wave field and the probed scene. The interaction between the EM wave field and the probed scene is nonlinear in essence; however, most of the mathematical formulations in practical EM imaging are linear, failing to capture the realistic underlying physical mechanism in a computationally tractable way.

Third, it is desirable to discover more realistic and richer low-dimensional models of underlying EM information. Up to this point, the low-dimensional models utilized in EM imaging are nearly simple sparse or compressible models; more realistic and richer models have not yet been fully investigated.

Finally, we are in the deluge of EM data coming from the continuously increasing demands on retrieving very detailed information of objects. Therefore, it is urgently necessary to develop efficient algorithms for treating massive measurements and high-dimensional variables.

4.2 Single-Frequency Multiple-Shot Compressive Metasurface Imager

As pointed out previously, one essential problem of compressive imagers is to construct the controllable masks or modulators as information relays; however, it was thought to be a challenging problem, especially in microwave frequencies and beyond. Nowadays, we can conclude from previous discussions that the reprogrammable coding metasurface can play this role. Nonetheless, there is still a fundamental but critical issue, namely the temporal-spatial resolution achievable, which is related to the sustainable throughput of the imager's illumination conditions, memory, and exposure time. In this section, we discuss the compressive metasurface imager proposed by Li et al.: single-frequency multiple-shot compressive imager with a one-bit reprogrammable coding metasurface [72]. Here, we mean by the multiple-shot, for instance, M-shot, that the probed object is illuminated M times by the metasurface-modulated EM source. It is clear that the value of M is the number of control coding patterns of reprogrammable metasurface and determines the number of measurements of compressive imaging in total. Unlike conventional CS imagers where the elements of random mask are individually manipulated, here the meta-atoms of reprogrammable metasurface are controlled in a column-row-wise manner.

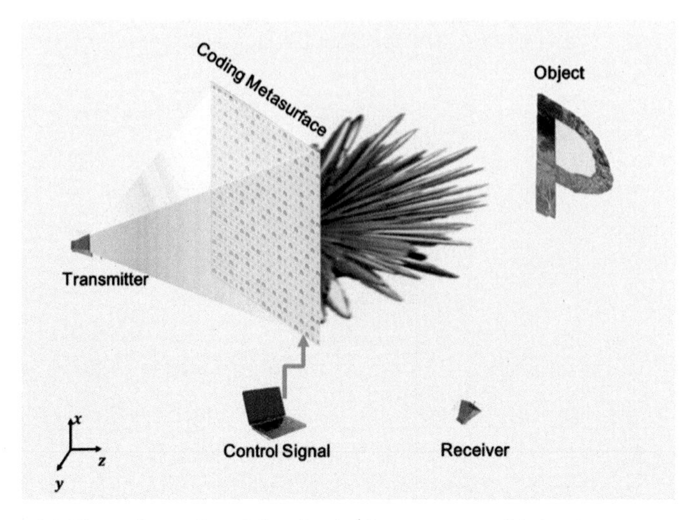

Figure 4.1 The schematic of the single-frequency multiple-shot compressive metasurface imager. This imager consists of a transmitter working with single frequency that launches an illumination wave. A one-bit coding metasurface is responsible for generating sequentially random masks for modulating the spatial wavefront emerging from the transmitter, and a single sensor is fixed somewhere for collecting the wave field scattered from the probed object.

Therefore, such CMI has a reduced data-acquisition time with improved obtainable spatial-temporal resolution, with respect to conventional CS imagers.

4.2.1 System Configuration

Here, we elaborate on the system configuration of a single-frequency multiple-shot compressive metasurface imager, as shown in **Fig. 4.1**, which consists of three major building parts: a single-frequency transmitter for launching the incident wave, a one-bit reprogrammable coding metasurface for modulating sequentially the spatial wavefront emanating from the transmitter, and a single sensor fixed for collecting the wave fields scattered from the probed object. In implementation, the one-bit reprogrammable coding metasurface is composed of 20×20 voltage-controllable ELC meta-atoms [77], shown in **Fig. 4.2a and b**. Each ELC meta-atom has the size of $6{\times}6$ mm^2 printed on the top surface of the FR4-substrate with a dielectric constant of 4.3 and a thickness of 0.2 mm, plotted in **Fig. 4.2b**. The ELC meta-atom is loaded with a pair of identical PIN-biased diodes (SMP 1320-079LF), and the one-bit states of "0" and "1" are controlled by the applied bias direct current (DC) voltage: when the biased voltage is on a high level (3.3 V), this pair of diodes are "ON"; when there is no biased voltage, the diodes are "OFF." The effective circuit models of the biased

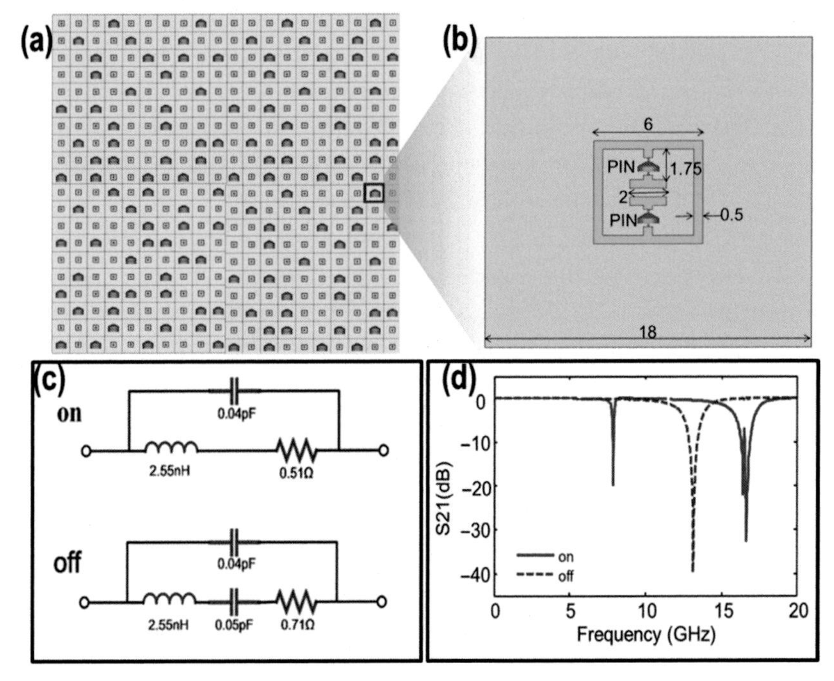

Figure 4.2 (a) The diagram of the one-bit reprogrammable coding metasurface composed of 20 × 20 voltage-controlled ELC particles. For visual purposes, the ELC particles coded with the ON state are highlighted in solid lines, and the others are not highlighted. **(b)** The configuration for the voltage-controlled ELC unit, which has a period of 18 mm, and a size of 6 × 6 mm^2. The ELC meta-atom is printed on a commercial printed circuit board FR4 with a relative permittivity of 4.3 and a thickness of 0.2 mm. **(c)** The effective circuit models of the biased diode at the ON and OFF states. **(d)** The S21 responses of the ELC meta-atom loaded with the PIN diode, implying that the ELC meta-atom behaves as a "1" element when the diode is on, and as a "0" element when the diode is off at the working frequency of 8.3 GHZ.

PIN diode at the ON and OFF states are illustrated in **Fig. 4.2c**. To show this more clearly, we plot the transmission responses (S_{21}) of the ELC particle loaded with the voltage-controlled PIN diodes in **Fig. 4.2d**. This figure clearly shows that, at the working frequency of 8.3 GHz, the ELC particle behaves as a "1" element when the diodes are ON, and as a "0" element when the diodes are OFF. Then a sequence of different random radiation patterns can be achieved by managing the applied voltages of the reprogrammable coding metasurface, which could provide adequate modes for our compressive measurements.

Although the proposed imager is an instance of the coded aperture imaging systems, it is different from the conventional CS-inspired imagers (e.g., the single-pixel camera [53] and a recent terahertz single-sensor imager [54]) where the elements of the random masks are manipulated in a pixel-wise manner. The controllable elements in this scheme are manipulated in a column-row-wise manner, which could greatly simplify the shutter control mechanism of the pixel-wise coded exposure. In particular, the reprogrammable one-bit coding metasurface composed of 20×20 meta-atoms is controlled by $20 + 20 = 40$ instead of $20 \times 20 = 400$ random binary sequences. Interestingly, it has been theoretically demonstrated, as detailed in what follows, that such column-row-wise coding metasurface gives rise to the performance of compressive imaging, which is comparable to that with the metasurface coded in the pixel-wise manner. As such, the reprogrammable coding metasurface is capable of producing quasi-random patterns in a very flexible and dynamic manner.

4.2.2 Imaging Model and Theoretical Guarantee

We turn to study the mathematical representation and imaging principle of the single-frequency M-shot compressive metasurface imager. Recall **Fig. 4.1**; we assume that the one-bit reprogrammable metasurface with the mth coded pattern, illuminated by an x-polarized plane wave, will approximately give rise to the radiation at $\boldsymbol{r}$:

$$E^{(m)}(\boldsymbol{r}) = \sum\nolimits_{n_x=1}^{N_x} \sum\nolimits_{n_y=1}^{N_y} \widetilde{A}^{(m)}_{n_x,n_y} g\left(\boldsymbol{r}, \boldsymbol{r}_{n_x,n_y}\right) \tag{4.3}$$

$m = 1, 2, \ldots, M,$

where $g\left(\boldsymbol{r}, \boldsymbol{r}_{n_x,n_y}\right) = \frac{\exp(jk_0|\boldsymbol{r}-\boldsymbol{r}_{n_x,n_y}|)}{4\pi|\boldsymbol{r}-\boldsymbol{r}_{n_x,n_y}|}$ denotes the 3D Green's function in free space, k_0 is the wavenumber, $\widetilde{A}^{(m)}_{n_x,n_y} = A^{(m)}_{n_x,n_y} \exp\left(j\varphi^{(m)}_{n_x,n_y}\right)$ is the x-polarized current with the amplitude $A^{(m)}_{n_x,n_y}$ and phase $\varphi^{(m)}_{n_x,n_y}$ induced on the (n_x, n_y)th meta-atom at $\boldsymbol{r}_{n_x,n_y}$ of the one-bit reprogrammable metasurface. Here, N_x (N_y) denotes the total number of meta-atoms of the one-bit reprogrammable coding metasurface along the x-direction (y-direction). Notice that $N_x = N_y = 32$ is used for our specific implementations. In **Eq. (4.3)** the summation is performed over all metasurface meta-atoms, where n_x and n_y denote the running indices of the reprogrammable coding metasurface along the x- and y-directions, respectively. In addition, under the assumption of Born scattering, the probed object with contrast function $O(\boldsymbol{r})$, falling into the investigation domain V, is illuminated by the wave field of **Eq. (4.3)** and gives rise to the scattering field at $\boldsymbol{r}_d$:

$$E^{(m)}(\boldsymbol{r}_d) = \sum_{n_x=1}^{N_x} \sum_{n_y=1}^{N_y} \widetilde{A}^{(m)}_{n_x,n_y} \int_V g\left(\boldsymbol{r}, \boldsymbol{r}_{n_x,n_y}\right) g(\boldsymbol{r}_d, \boldsymbol{r}) O(\boldsymbol{r}) d\boldsymbol{r} \tag{4.4}$$

$m = 1, 2, \ldots, M.$

After introducing the following function,

$$\widetilde{O}_{n_x,n_y} = \int_V g(\boldsymbol{r}, \boldsymbol{r}_{n_x,n_y}) g(\boldsymbol{r}_d, \boldsymbol{r}) O(\boldsymbol{r}) d\boldsymbol{r}, \tag{4.5}$$

where $n_x = 1, 2, \ldots, N_x$, $n_y = 1, 2, \ldots, N_y$,
Equation (4.4) becomes

$$E^{(m)}(\boldsymbol{r}_d) = \sum_{n_x=1}^{N_x} \sum_{n_y=1}^{N_y} \widetilde{A}^{(m)}_{n_x,n_y} \widetilde{O}_{n_x,n_y} \tag{4.6}$$

$m = 1, 2, \ldots, M.$

By applying the far-field approximation, Eq. (4.6) can be expressed as

$$\widetilde{O}_{n_x,n_y} = \left(\frac{1}{4\pi}\right)^2 \frac{\exp\left(jk_0 r_{n_x,n_y}\right)}{r_{n_x,n_y}} \frac{\exp(jk_0 r_d)}{r_d} \int_V \exp\left(-j\left(\boldsymbol{k}_{n_x,n_y} + \boldsymbol{k}_d\right)\Delta \boldsymbol{r}\right) O(\boldsymbol{r}) d\boldsymbol{r} \tag{4.7}$$

where $\boldsymbol{k}_{n_x,n_y} = k_0 \frac{\boldsymbol{r}_{nx,ny}}{r_{nx,ny}}$ and $\boldsymbol{k}_d = k_0 \frac{\boldsymbol{r}_d}{r_d}$.

Equation (4.7) indicates that $\widetilde{O}_{n_x,n_y}$ corresponds to the two-dimensional discrete Fourier transform of $O(\boldsymbol{r})$. Notice that the spatial bandwidth of $O(\boldsymbol{r})$ is limited by the maximum value of $|\boldsymbol{k}_{n_x,n_y} + \boldsymbol{k}_d|$ and is determined by the maximum size of the coding metasurface. In other words, it can be deduced that the achievable resolution on $O(\boldsymbol{r})$ is in the order of $O(\lambda R/D)$[72], where λ is the operating wavelength, R is the observation distance, and D is the maximum size of the coded aperture.

In the context of computational imaging, **Eq. (4.7)** can be reformulated into the following compact form:

$$E^{(m)} = \widetilde{\boldsymbol{A}}^{(m)}, \widetilde{\boldsymbol{O}} \quad m = 1, 2, \ldots, M, \tag{4.8}$$

where the symbol $\langle . \rangle$ denotes the matrix inner product, the matrix $\widetilde{\boldsymbol{A}}^{(m)}$ has the size of $N = N_x \times N_y$ with entries of $\widetilde{A}^{(m)}_{n_x,n_y}$, and the matrix $\widetilde{\boldsymbol{O}}$ with the size of N is populated by $\widetilde{O}_{n_x,n_y}$. **Equation (4.8)** reveals that the resulting computational imaging problem consists of retrieving N unknowns $\{\widetilde{O}_{n_x,n_y}\}$ from the M measurements $\{\boldsymbol{E}^{(m)}(\boldsymbol{r}_d)\}$. Typically, **Eq. (4.8)** has no unique solution if $N > M$ due to its intrinsic ill-posedness. To overcome this difficulty, we pursue a sparsity-regularized solution to **Eq. (4.8)** since we believe that the probed object

$\widetilde{\boldsymbol{O}}$ has a low-dimensional representation in a certain transform domain denoted by $\boldsymbol{\Psi}$, namely $\boldsymbol{\Psi}\left(\widetilde{\boldsymbol{O}}\right)$ being sparse. Therefore, the solution to **Eq. (4.8)** could be achieved by solving the following sparsity-regularized optimization problem:

$$min_{,O}\left[\frac{1}{2}\sum_{m=1}^{M}\left(E^{(m)}-\langle\widetilde{\boldsymbol{A}}^{(m)},\widetilde{\boldsymbol{O}}\rangle\right)^{2}+\gamma||\boldsymbol{\Psi}\left(\widetilde{\boldsymbol{O}}\right)||_{1}\right], \quad (4.9)$$

where γ is a balancing factor to trade off the data fidelity and the sparsity prior.

As pointed out previously, the coded elements of random mask are conventionally controlled in a pixel-wise fashion, which gives rise to too much exposure time and thus limits the temporal and spatial resolutions. To address this issue, we propose the column-row-wise coding metasurface, in which the time-varying random modulation of $N_x \times N_y$ meta-atoms is controlled by $N_x + N_y$ rather than $N_x \times N_y$ random binary sequences. Specifically, the N_x-length random binary sequences of $\{0,\ 1\}^{N_x}$ (denoted by $\boldsymbol{r} = [r_1, r_2, \ldots, r_{N_x}]$) and the N_y-length random binary sequences of $\{0,\ 1\}^{N_y}$ (denoted by $\boldsymbol{c} = [c_1, c, \ldots, c_{N_y}]$) are used to control the row and column pixels, respectively. The row and column binary control signals jointly produce the binary random coded exposure sequence at the pixel location of (n_x, n_y). Thus, one random realization of a coded pattern reads $\widetilde{\boldsymbol{A}} = \boldsymbol{r}^T\boldsymbol{c}$ up to a constant multiplicative factor. In this design, only $N_x + N_y$ control signals are needed to achieve the randomly coded exposures with $N_x \times N_y$ pixels, which could drastically reduce the complexity and increase the filling factor. We demonstrated that our imager based on the one-bit column-row-wise coding metasurface has a theoretical guarantee on successful recovery of a sparse or compressible object from its reduced measurements by solving a sparsity-regularized convex optimization problem, which is comparable to that by the conventional pixel-wise coded imaging system. The conclusion is summarized in **Theorem 2** [72].

THEOREM 2

With *M*, *N*, and *S* defined as previously, an *S*-sparse *N*-length signal can be accurately retrieved with a probability not less than $2\exp\left(-C(\log \boldsymbol{S} \log \boldsymbol{N})^2\right)$, provided that the number of measurements with $\boldsymbol{M} \geq \boldsymbol{C} * \boldsymbol{S} \log \boldsymbol{N}/\boldsymbol{S}$ is up to a polynomial logarithm factor, where *C* is a constant depending only on *S*.

Proof of Theorem 2

The proof here resembles used a powerful theorem obtained in [73]. For the readers' convenience, we repeat this theorem here to facilitate our proof.

THEOREM 3

Let $\mathcal{A} \in \mathbb{C}^{M \times N}$ be a set of complex-valued matrices, and let ε be a random vector whose entries are i.i.d. zero-mean, unit variance random variables with the sub-Gaussian norm τ. Set $d_F(\mathcal{A}) = \sup_{A \in \mathcal{A}} ||A||_F$, $d_2(\mathcal{A}) = \sup_{A \in \mathcal{A}} ||A||_2$,

$E_1 = \gamma_2(\mathcal{A}, ||\cdot||_2)\{\gamma_2(\mathcal{A}, ||\cdot||_2) + d_F(\mathcal{A})\} + d_F(\mathcal{A})d_2(\mathcal{A})$,

$E_2 = d_2(\mathcal{A})\gamma_2(\mathcal{A}, ||\cdot||_2) + d_F(\mathcal{A})$, and $E_3 = d_2^2(\mathcal{A})$, then, for $t > 0$, it holds that

$$\log P\left\{ \sup_{A \in \mathcal{A}} \left| ||A\varepsilon||_2^2 - \mathbb{E}||A\varepsilon||_2^2 \right| \geq_\tau E_1 + t \right\} \leq_\tau \min\left(\frac{t^2}{E_2^2}, \frac{t}{E_3} \right).$$

To prove **Theorem 2**, we assume that $||X||_F = 1$ without loss of generality, and then reformulate **Eq. (4.8)** as

$$\langle \mathbf{X}, \mathbf{\Phi}_i \rangle = Trace\left(\boldsymbol{X}^T \boldsymbol{c}_i \boldsymbol{r}_i^T \right) = \boldsymbol{r}_i^T \boldsymbol{X}^T \boldsymbol{c}_i, \tag{4.10}$$

which can be furthermore expressed as

$$\frac{1}{\sqrt{M}} \mathbf{\Phi}_i,\ \mathbf{X} = d_i \quad m = 1, 2, \ldots, M \tag{4.11}$$

or

$$\begin{bmatrix} (\boldsymbol{X}\boldsymbol{r}_i)^T & & \\ & \ddots & \\ & & (\boldsymbol{X}\boldsymbol{r}_M)^T \end{bmatrix} \begin{bmatrix} \boldsymbol{c}_1 \\ \vdots \\ \boldsymbol{c}_M \end{bmatrix} = \begin{bmatrix} d_1 \\ \vdots \\ d_M \end{bmatrix}. \tag{4.12}$$

It is readily shown that

$$d_F(\mathcal{A}) = \sup_{A \in \mathcal{A}} ||A||_F = \frac{1}{M} \sup_{A \in \mathcal{A}} \sum_{i=1}^{M} ||\boldsymbol{X}\boldsymbol{r}_i||_2^2 = 1 \tag{4.13}$$

and

$$d_2(\mathcal{A}) \leq \sqrt{S/M}. \tag{4.14}$$

In **Eq. (4.14)**, $\sum_{i=1}^{M} ||\boldsymbol{X}\boldsymbol{r}_i||_2^2 = Trace \sum_{i=1}^{M} \boldsymbol{X}\boldsymbol{r}_i \boldsymbol{r}_i^T \boldsymbol{X}^T = MTrace\left(\boldsymbol{X}\boldsymbol{X}^T\right) = M$, where $||\boldsymbol{X}||_F = 1$ is used in the last equation, and $\boldsymbol{r}$ is an i.i.d. zero-mean and unit variance random vector. In **Eq. (4.14)**, we obtain the following conclusion:

$$
\begin{aligned}
d_2(\mathcal{A}) &= \sup_{A\in\mathcal{A}} ||A||_2 = \sup_{A\in\mathcal{A}} \sqrt{||AA^T||_2} \\
&= \frac{1}{\sqrt{M}} \sup_{A\in\mathcal{A}} max_i ||Xr_i||_2 \\
&\leq \frac{||r_i||_\infty}{\sqrt{M}} ||X||_1 \\
&\leq \sqrt{\frac{S}{M}} ||r_i||_\infty ||X||_F \\
&= \sqrt{\frac{S}{M}} ||r_i||_\infty .
\end{aligned}
$$

On the other hand, the upper bound of $\gamma_2(\mathcal{A}, ||\cdot||_2)$ can be estimated as

$$
\gamma_2(\mathcal{A}, ||\cdot||_2) \leq C\sqrt{\frac{S}{M}} \log S \log N. \tag{4.15}
$$

Using Eqs. (4.13)–(4.15) in Theorem 2 immediately leads to the conclusion that $\mathrm{P}\left(\delta_S \geq \hat{\delta}\right) \leq 2\exp\left(-C(\log S \log N)^2\right)$ holds for $M \geq C\hat{\delta}^{-2} S(\log S \log N)^2$, which implies that the measurement operator $\mathcal{A}$ satisfies the so-called S-RIP with the isometry constant $0 < \delta_S < 1$ with a probability not less than $2\exp\left(-C(\log S \log N)^2\right)$, provided that the number of measurements $M \geq C\hat{\delta}^{-2} S(\log S \log N)^2$, where C is a constant only depending on δ_S. As a consequence, **Theorem 2** can be obtained immediately.

A set of Monte Carlo simulation results are provided to verify **Theorem 2**, where the signals of dimension $N = N_x \times N_y = 32 \times 32$ whose nonzero entries are drawn i.i.d. from a standard normal distribution and are located on a support set drawn uniformly at random. We varied the number of random masks from $M = 8$ to $M = 1024$ with the step of size 8. Among all 100 trials run for each pair of M and S, the ones that yield relative errors no more than 10^{-3} are counted as successful reconstruction. For comparison, the meta-atoms of metasurface that are randomly controlled in the pixel-wise manner are investigated as well. **Figure 4.3a and b** shows the phase transition diagrams of the estimation produced by **Eq. (4.9)** in terms of the number of masks M and S for the proposed column-row-wise and conventional pixel-wise coded masks, respectively. In these figures, the x- and y-axes correspond to the values of M and S, respectively. We observe that the phase transition boundaries are almost linear in both models, agreeing with the relation between M and S suggested by

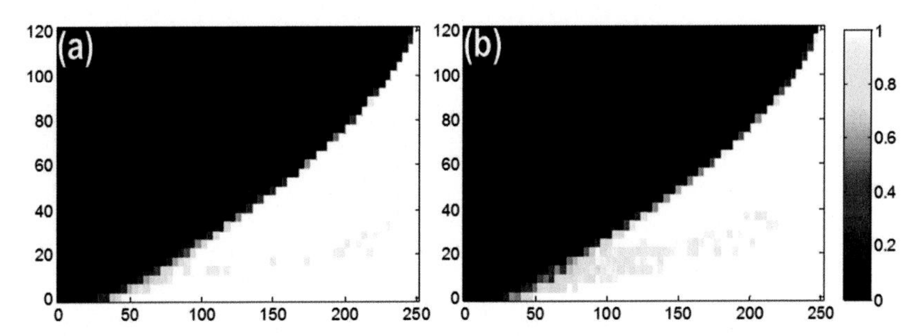

Figure 4.3 The phase transition diagram of l1-minimization in terms of M, the number of masks, and S, the l0-norm of the signals with dimension $L = 1024$. **(a)** The column-row-wise coded system. **(b)** The pixel-wise coded system. In these figures, the x-axis denotes M while the y-axis corresponds to S.

Theorem 2. Although the phase transition boundary for the column-row-wise model is slightly lower (worse) than that for the pixel-wise model, the difference is not significant.

4.2.3 Results

A set of proof-of-concept experiments were conducted to verify the performance of the imager discussed earlier. For this purpose, we fabricated a sample of a one-bit reprogrammable coding metasurface, as shown in **Fig. 4.4a**. In our experiments, a vector network analyzer (VNA, Agilent E5071C) was used to acquire the response data by measuring the transmission coefficients (S21). More specifically, a pair of horn antennas were connected to two ports of the VNA through two 4-m-long 50-Ω coaxial cables: one was used for launching the incident wave, and the other for collecting the response data emanated from the probed object, as shown in **Fig. 4.4b**.

As for the control of the reprogrammable coding metasurface, the biased voltages of metasurface meta-atoms could be digitally controlled by toggling different triggers in both column and row distributions. In this way, the "ON" and "OFF" states of the biased PIN diodes can be controlled, and thereby the required "0" or "1" state of each metasurface particle can be realized. In this way, we generated 1000 random radiation patterns in total, and we plot three of them in **Fig. 4.4d.** These radiation patterns were obtained by using the so-called near-field scanning technique along with the near-far-field transform. Then, two sets of imaging results are presented in **Fig. 4.4**. As done in the numerical simulations, the "P" and "K"-shaped metallic objects are considered in the experiments, shown in **Fig. 4.4c**. The reconstruction results for "P" ("K") are provided in **Fig. 4.4e–g** (**Fig. 4.4h–j**), considering different numbers of measurement $M = 200$, 400, and

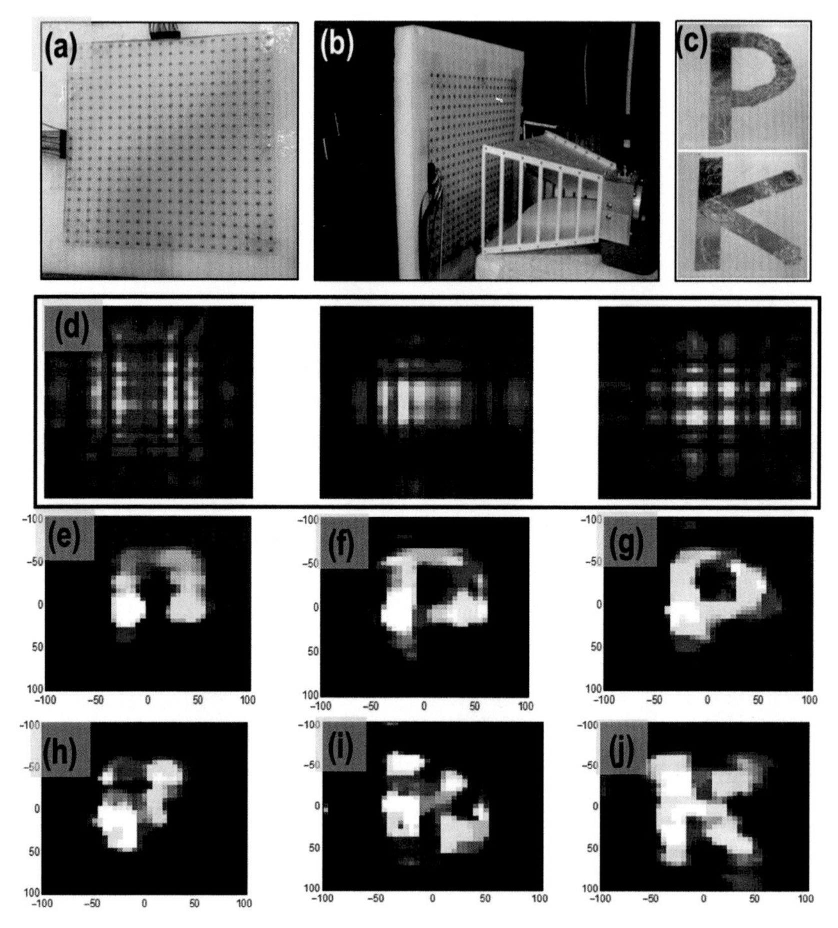

Figure 4.4 (a) The fabricated sample of the column-row-wise coding metasurface. **(b)** The experimental single-sensor imaging system based on the 1-bit programmable coding metasurface. **(c)** The "P"- and "K"-shaped metallic objects for the imaging test. **(d)** Three samples of the radiation patterns of the metasurface encoded with the controlling signals. **(e–g)** The measured imaging results for the "P"-shaped metallic object with different measurement numbers M = 200, 400, and 600, respectively. **(h–j)** The measured imaging results for the "K"-shaped metallic object with different measurement numbers M = 200, 400, and 600, respectively.

600, respectively. The experimental imaging results clearly validate the feasibility of the proposed single-frequency multiple-shot compressive metasurface imager based on the one-bit column-row-wise coding.

To summarize, we discussed the single-frequency multiple-shot compressive metasurface imager based on the one-bit column-row-wise reprogrammable coding metasurface. From the preceding discussion, we see that such an imager has two advantages over existing compressive imagers: (1) It avoids the frequency dispersion of target without the frequency agility; (2) it encodes the reprogrammable metasurface in the column-row-wise manner instead of the pixel-wise manner to reduce drastically the data acquisition time, resulting in an improved temporal-spatial resolution. Interestingly, this simple one-bit column-row-wise coding metasurface has a theoretical guarantee that ensures the required measurement number is comparable to that for the conventional pixel-wise encoded masks while maintaining nearly the same imaging quality. This new imaging system could be readily extended to other frequencies and beyond.

4.3 Wideband Single-Shot Compressive Imager

Real-time high-resolution (including super-resolution) imaging with low-cost hardware is a long-sought-after goal in various imaging applications. We discuss the broadband single-shot metasurface compressive imager for high/super-resolution imaging by using a spatio-temporal dispersive metasurface and computational imaging algorithm. For this purpose, the metasurface used here is designed with the property of spatio-temporal dispersion [71], which is capable of converting the spatial information of the probed object into 1D time- or frequency-dependent signal. As a consequence, the 3D high-resolution or super-resolution spatial information of the target can be sufficiently retrieved by using a single wideband sensor fixed in the far-field region of the target and implementing a simple back-propagation reconstruction algorithm [74–76]. Of course, the image quality can be further improved by performing a feature-enhanced reconstruction algorithm, and the imaging resolution achievable is controlled by adjusting the distance between the target and the metasurface. Specifically, when the target is in the vicinity of the metasurface, the super-resolution imaging can be achieved, since the evanescent-wave component carrying the subwavelength information of target can be captured by the spatial-temporal-dispersive metasurface and transferred to the far-field sensor. Such imaging methodology enables us to get the high-/super-resolution images with real-time data acquisition, but without using expensive hardware.

4.3.1 System Configuration and EM Theory

System Configuration The wideband single-shot compressive metasurface imager is conceptually illustrated in **Fig. 4.5a**, where a wideband EM source along with a horn antenna with a frequency band of 7–10.2 GHz is deployed to illuminate the target of interest. A single sensor (another horn antenna) fixed at $\boldsymbol{r}_d$ is utilized to acquire the responses of the single-shot illumination scattered from the target, and

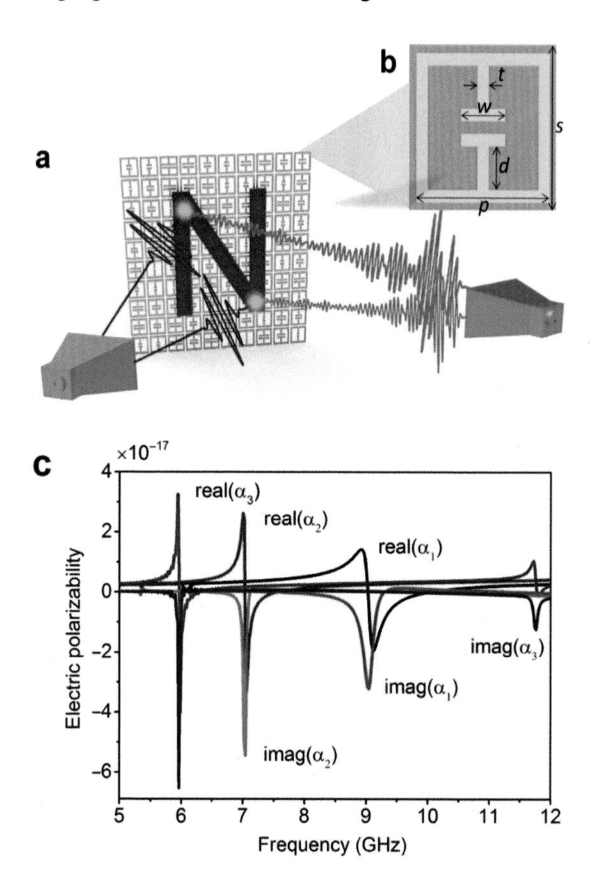

Figure 4.5 (a) The single-shot and single-sensor far-field imaging system based on a metasurface. **(b)** The geometry of unit cell. **(c)** The electric polarizabilities of unit cells retrieved from the method of dipole approximation when $w = 1$ mm, 2.6 mm, and 4.0 mm, respectively. [71]

a frequency-dispersive metasurface is situated in front of the probed object on the plane of $z = 0$. A vector network analyzer (VNA, Agilent E5071C) was utilized for the experimental far-field imaging, and the experiment was carried out in a microwave anechoic chamber with the size of 2 m × 2 m × 2 m. The S_{21} parameters (i.e., transmission coefficients) were measured as the data input for the imaging reconstruction algorithms. The calibration of the mapping matrix $\boldsymbol{A}$ is accomplished by the point-to-point raster scanning method, namely sequentially moving a small object (a small Chinese coin) and recording corresponding S_{21} responses. To suppress the measurement noise level, the average number and filter bandwidth were set to be 10 and 10 kHz in VNA, respectively.

The frequency-dispersive metasurface is composed of an array of 30 × 30 passive ELC meta-atoms with random resonant frequencies, as shown in

Fig. 4.5. It is fabricated using a commercial FR4 substrate with a thickness of 0.5 mm, a dielectric constant of 4.6, and a loss tangent of 0.002. Other parameters of ELC meta-atoms are set as follows: S = 6.67 mm, p = 6 mm, t = 0.5 mm, d = 1.75 mm. The resonant frequency of a meta-atom can be controlled with w, and w was selected randomly from 12 choices of 0.6 mm, 1 mm, 1.4 mm, 1.8 mm, 2.2 mm, 2.6 mm, 2.8 mm, 3.2 mm, 3.6 mm, 4.0 mm, 4.4 mm, and 4.8 mm. To model the EM response of metasurface theoretically, the so-called coupled-dipole method (CDM) was explored, in which each metasurface meta-atom is modeled as an electric polarizability tensor. The dominant components (i.e., z-components) of electric polarizability tensors for three typical values of w (1 mm, 2.6 mm, and 4.0 mm) are plotted in **Fig. 4.5b**, which are extracted by performing the commercial software, CST Microwave Studio 2014.

As argued earlier, the wide-band single-shot compressive metasurface imager relies on the utilization of a spatio-temporal dispersive metasurface, which is responsible for converting sufficiently the 3D spatial information of the target with the reflectivity $O(x, y, z)$ to the 1D time-dependent response $s(t)$ received by the sensor fixed at $\boldsymbol{r}_d$. To see clearly the physical mechanism behind it, a set of full-wave numerical simulations were conducted and the results are shown in Fig. 4.6, in which the metasurface is illuminated by a transverse-electric (TE) polarized plane wave $\boldsymbol{E}_{in} = \boldsymbol{E}_0 \exp(i\boldsymbol{k}_{in} \cdot \boldsymbol{r})$. Here, $\boldsymbol{k}_{in} = \left(k_{inx}, k_{iny}, k_{inz}\right)$, $k_{inz} = \sqrt{k_0^2 - k_{inx}^2 - k_{iny}^2}$, and k_0 is the free-space wavenumber. The case of $k_{inp} = \sqrt{k_{inx}^2 + k_{iny}^2} > k_0$ corresponds to the evanescent wave illumination, while the case of $k_{inp} < k_0$ corresponds to the propagating wave. As mentioned earlier, the coupled-dipole method was performed to calculate the EM response of the metasurface. The calculated amplitude of scattered electric fields at $\boldsymbol{r}_d$ as functions of operational frequencies and transverse wavenumbers are plotted in **Fig. 4.6**. Here, the horizontal axis denotes the transverse wavenumber normalized by k_0 (the free-space wavenumber at 10 GHz), while the vertical axis denotes the operational frequency normalized by ω_p (the angular frequency at 10 GHz). This map clearly demonstrates that different wavenumber components can be transferred into the temporal domain successfully after passing through the metasurface with substantially high efficiency.

4.3.2 Results

Far-field imaging beyond the diffraction limit is desirable in various areas of research. In this topic three research streams have gained intensive attention in the past decade, including the super-oscillation [28–31], the resonant structure

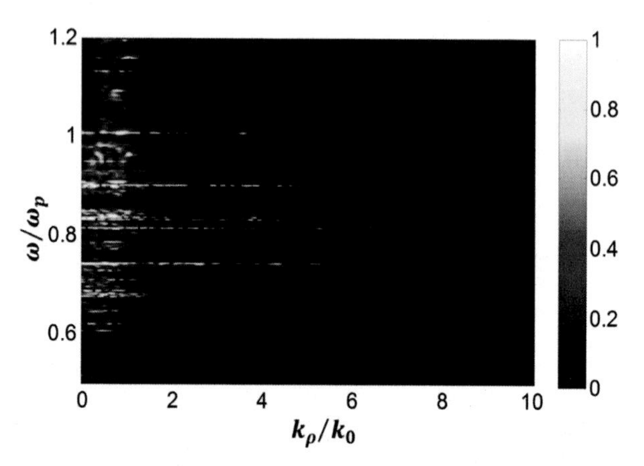

Figure 4.6 The amplitudes of electric fields recorded by a sensor in the far-field region, in which the horizontal axis represents k_x normalized to k_0 (the free-space wavenumber at 10 GHz), while the vertical axis represents the frequency normalized to ω_p.

[23–24], and others [32–33]. The essential requirement of such techniques is the sensor array or mechanical movement, which typically requires expensive hardware and time-consuming data acquisition. As demonstrated in **Fig. 4.6**, the spatial-temporal dispersive metasurface is capable of converting the evanescent waves emerging from the target in the small vicinity of metasurface into the propagating waves, allowing the reconstruction of subwavelength-scale information of the object registered in the one-dimensional temporal data acquired by the single sensor fixed in the far fields. To verify this claim, we here present a set of numerical results selected from [71]. In simulations, the metasurface with the 10×10 meta-atoms with random resonant frequencies was considered, which was placed in the vicinity of the probed target; that is, the speciman was located at a distance of 4 mm away from the metasurface. The specimen under investigation was a capital character "N" made up of dielectric bars with a refractive index of 4.3 and a rectangular cross section of 0.5 mm $\times$ 6.67 mm. We voluntarily opted for a low-refractive-index contrast, standing for the soft-matter object. In this study, the white noise with a signal-to-noise ratio (SNR) of 30 dB is added to the CST simulation data. In **Fig. 4.7b and c**, we plot the imaging results from the traditional minimum least-square reconstruction algorithm and the iteratively reweighting reconstruction algorithm, respectively, which clearly show that the two objects, separated by a distance of 6.67 mm (about 0.2λ), can be clearly resolved from far-field single-sensor measurements. We underline here that such super-resolution imaging works in real time, since it only requires a single-shot illumination

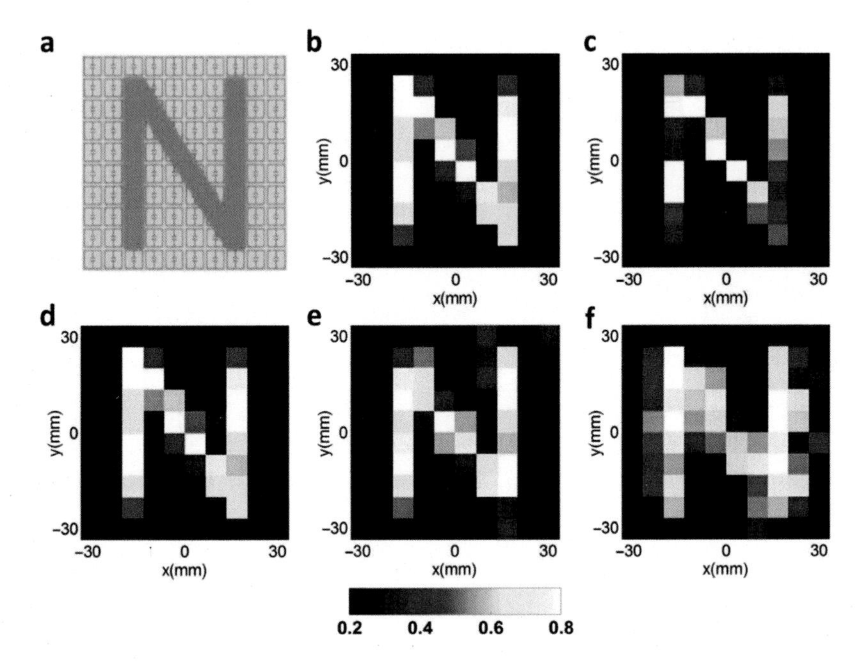

Figure 4.7 Far-field subwavelength imaging results, in which we emit the single broadband illumination using a horn antenna and receive the time-dependent scattered fields in the far-field region by another single horn antenna. **(a)** An N-shaped dielectric object located near the metasurface. **(b)** The imaging results using the traditional least square algorithm. **(c)** The imaging results using the sparse-aware reconstruction algorithm. Here, an imaging resolution around λ/5 has been achieved. In **(d)–(f)**, the distance between the object and metasurface is 4 mm, 10 mm, and 20 mm, respectively.

and single-sensor data acquisition from the far field, removing the harsh requirements of the use of a sensor array, mechanical movement, and near-field measurements.

The concept of the broadband single-shot compressive metasurface imager relies on a critical point: the temporal-spatial metasurface is responsible for encoding the 3D spatial structures of the probed object into the 1D temporal signals. As a consequence, the imaging resolution depends largely on the distance between the metasurface and the object under investigation. To demonstrate this issue clearly, we here present a set of CST-based simulation results selected from [71], where the imaged object is the same as that in **Fig. 4.7a**. **Figure 4.7d–f** shows three reconstructed images of the object for three representative distances between the metasurface and object, in which the simulation parameters are the same as those in **Fig. 4.7b**, and the standard minimum least-square reconstruction algorithm is used for data processing. From these results, we observe that the imaging quality will become worse as the distance of the object away from the metasurface grows up, since for the relatively bigger distance, the evanescent wave carrying the finer structures of the probed object will be lost before arriving at the metasurface. Here, we would like to say that the advanced reconstruction algorithm could be incorporated into the imaging procedure to improve the imaging quality substantially[20].

To summarize, we have presented a broadband single-shot and single-sensor system for high-resolution and subwavelength imaging, in combination with the metasurface that can convert the spatial information of the probed object into the temporal data measured by the single sensor [80–85]. We can find that such an imager has the unique ability to achieve high-resolution and super-resolution images in a real-time manner without using a sensor array as well as mechanical movement and near-field scanning of a sensor. We expect that such an imaging scheme could make breakthroughs in imaging technologies in the microwave, terahertz, optical, and ultrasound regimes.

5 Machine-Learning Metasurface Imager

The most important but challenging problem for most modern sensing systems is to efficiently deal with the high-dimensional data or "data crisis." Fortunately, the Johnson–Lindenstrauss lemma states that the structured high-dimensional data can be projected into a low-dimensional space with nearly negligible information loss through a properly designed linear transform, which is the theoretical foundation of compressive imagers discussed in the previous section. However, it is a formidable challenge for conventional imagers to generate the radiation patterns required by machine-learning-driven *meaningful* measurement modes in real time, and suffers from the massive data collection and expensive data processing. To tackle this challenge, recently Li et al. proposed the machine-learning metasurface imager on the basis of an inexpensive two-bit reprogrammable coding metasurface [86]. It has been demonstrated that, after being trained with machine learning techniques, such an imager can be designed with the meaningful measurement modes required by machine learning techniques, which enables us to realize real-time and high-quality imaging and recognition from remarkably reduced measurement with nearly negligible digital computation. We expect that such a sensing technique will have great potential to impact imaging applications in the microwave, millimeter wave, and terahertz regimes and beyond.

5.1 Sensing Principle

Generally speaking, EM imaging consists of retrieving a scene from its measurements of scattered EM responses, which, under the Born or single-scattering approximation [87–88], is usually attributed to establish a linear model bridging the measured return signal and the scene. Formally, this problem can be represented as $\boldsymbol{y} = \mathbf{H}\boldsymbol{x} + \boldsymbol{n}$, where $\boldsymbol{y} \in \mathbb{C}^M$ is the M-length complex-valued measurements, $\mathbf{H}$ denotes the complex-valued sensing matrix, $\boldsymbol{x} \in \mathbb{C}^N$ is the N-length complex-valued vector representation of the probed scene, and $\boldsymbol{n}$ is the measurement noise. Under the Born approximation, the entry of the measurement matrix (H_{ij}) is simply proportional to the fields radiated by the transmitter and receiver antennas at a given point in the scene $\boldsymbol{r}_j$: $H_{ij} \propto \boldsymbol{E}_i^T(\boldsymbol{r}_j) \cdot \boldsymbol{E}_i^R(\boldsymbol{r}_j)$, where $\boldsymbol{E}_i^T(\boldsymbol{r}_j)$ and $\boldsymbol{E}_i^R(\boldsymbol{r}_j)$ represent the radiation and receiving patterns, respectively, related to the one-bit reprogrammable coding metasurface. Each row of the measurement matrix corresponds to a measurement mode, and hence the number of rows equals the number of measurements, and the number of columns equals the number of voxels in the probed scene. More specifically, the number of measurement modes is determined by the control coding patterns of the reprogrammable coding metasurface. The imaging solution reads $\hat{\boldsymbol{x}} = \boldsymbol{H}^+\boldsymbol{y}$, where $\boldsymbol{H}^+$ is the pseudo-inverse matrix of $\mathbf{H}$.

For most modern sensing systems, the most important yet challenging problem is dealing with the high-dimensional data. From the perspective of machine learning, the reduced measurements can be regarded as linear embedding of the probed scene in the low-dimensional space. The well-known Johnson–Lindenstrauss lemma states that structured high-dimensional data could be linearly projected into a low-dimensional feature space with nearly negligible information loss through a properly designed linear transform [89]. In other words, the essential information of high-dimensional data can be efficiently retrieved from its remarkably reduced measurements in most practical settings. Given a sample $\boldsymbol{x} \in \mathcal{X}$, where $\mathcal{X}$ is a cloud of Q points in an N-dimensional Euclidean space, the low-dimensional linear embedding method consists of finding a projection matrix $\boldsymbol{H}$ such that the M-dimensional projection $\boldsymbol{y} = \boldsymbol{H}\boldsymbol{x}$ has as small a loss of intrinsic information as possible compared to $\boldsymbol{x}$, where $M \ll N$. An approach that has attracted intensive attention in the past decade is the random projection, in which the entries of $\boldsymbol{H}$ are drawn from independent random numbers. The random approach imposes no restrictions on the nature of the object to be reconstructed. On the contrary, PCA is a prior data-aware embedding technique, where each row of $\boldsymbol{H}$ (i.e., the measurement mode) is trained over the many training samples available [5–6]. In this way, when a set of prior knowledge on the scene under investigation is available, the PCA

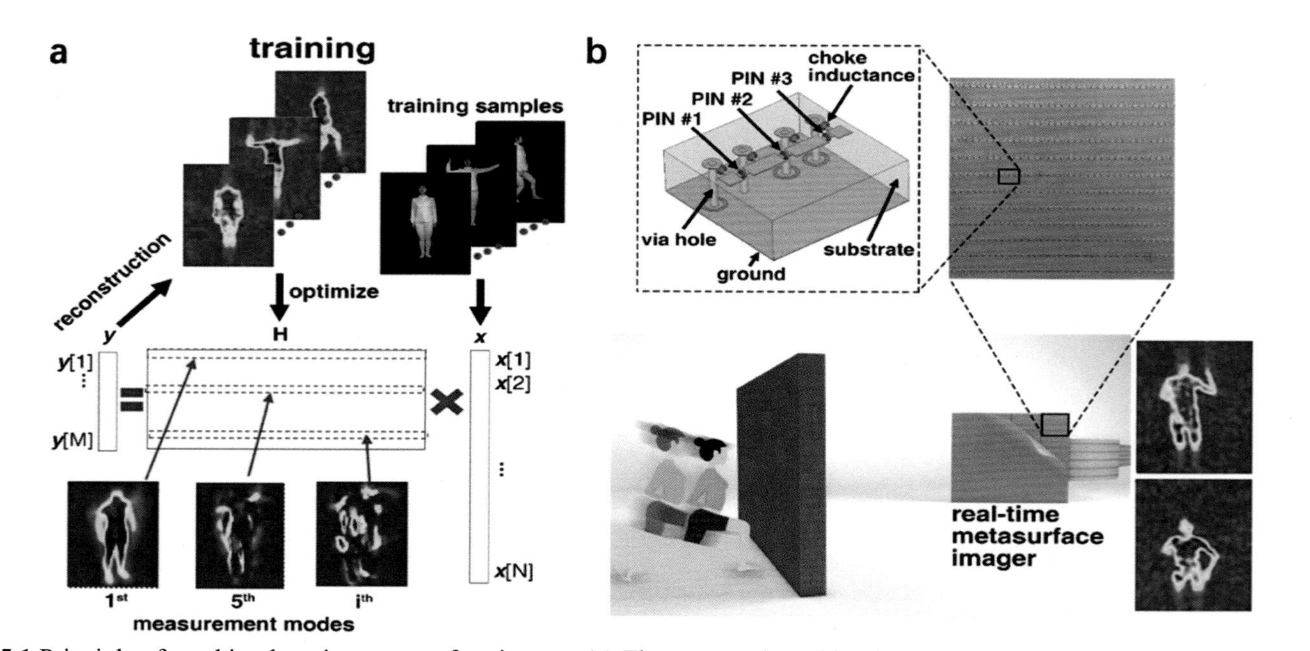

Figure 5.1 Principle of machine-learning metasurface imager. **(a)** The proposed machine-learning metasurface imager is optimized by the training samples. The scene $\boldsymbol{x}$ is compressed through the matrix $\mathbf{H}$: $\boldsymbol{y} = \mathbf{H}\boldsymbol{x}$. By minimizing the difference between the reconstructed scenes and the original scenes, the optimized matrix $\mathbf{H}$ is determined. **(b)** Photo of a two-bit reprogrammable metasurface, with the sketch of a cell on the left. [86]

approach enables the design of efficient measurement matrices, allowing the number of measurements to be limited compared to a purely random system.

Now, we turn to discuss the object recognition from the compressed measurements. Assume that there are K classes of labeled samples, and the center of the kth class is $\widetilde{\boldsymbol{y}}_k = \frac{1}{C_k}\sum_{t=1}^{C_k}\mathbf{H}\boldsymbol{x}_k^t, k = 1, 2, \ldots, K$, where C_k is the total number of samples in the kth class. For a given image $\boldsymbol{x}$ and its reduced measurements $\boldsymbol{y} \approx \boldsymbol{Hx}$, the object recognition consists of comparing $\boldsymbol{y}$ with the stored vectors $\widetilde{\boldsymbol{y}}_k(k = 1, 2, \ldots, K)$ and selecting the one with the closest match, which is classified mathematically with the minimum Euclidean distance $\boldsymbol{c} = \mathrm{argmin}_k\left(\|\boldsymbol{y}' - \widetilde{\boldsymbol{y}}_k\|_2^2\right)$. This concept is implemented in the machine-learning metasurface imager, from which a real-time object recognition system is developed. Once again, two aforementioned machine-learning techniques (random projection [91] and PCA [89–90]) are applied to train the proposed machine-learning imager.

5.2 System Design

This section discusses the system configuration of the machine-learning metasurface imager working at around 3.2 GHz. As shown in **Fig. 5.2**, here the system consists of four main building parts: a host computer, a FPGA microcontroller, a two-bit reflection-type reprogrammable coding metasurface, and a vector network analyzer (VNA). The whole reprogrammable metasurface consists of 3×4 metasurface panels, and each panel consists of 8×8 electrically controllable macro meta-atoms. Each macro meta-atom can be independently controlled, and an incident illumination on metasurface with different coding sequences will result in different backward-scattered fields. The metasurface is controlled by a Cycone-4 FPGA with a clock frequency of 300 MHz. To reduce the complexity of the writing command and controlling system, three SN74LV595A 8-bit shift registers are utilized to control the rest of the PIN diodes sequentially. When the PIN diodes are controlled in parallel by a FPGA, three operation cycles are needed after compiling, and thus the total image switching time is around tens of nanoseconds. The SPI (Serial Programming Interface Bus)-controlled command and data between the host computer and the FPGA, and VNA are communicated via Ethernet (IEEE 488.2). More details about the metasurface used here can be found in Section 2.4.

The machine-learning metasurface imager relies on a two-bit reprogrammable coding metasurface), which is capable of producing the physically meaningful measurement modes needed by linear machine learning with high accuracy in real time. In other words, the reprogrammable coding measurface

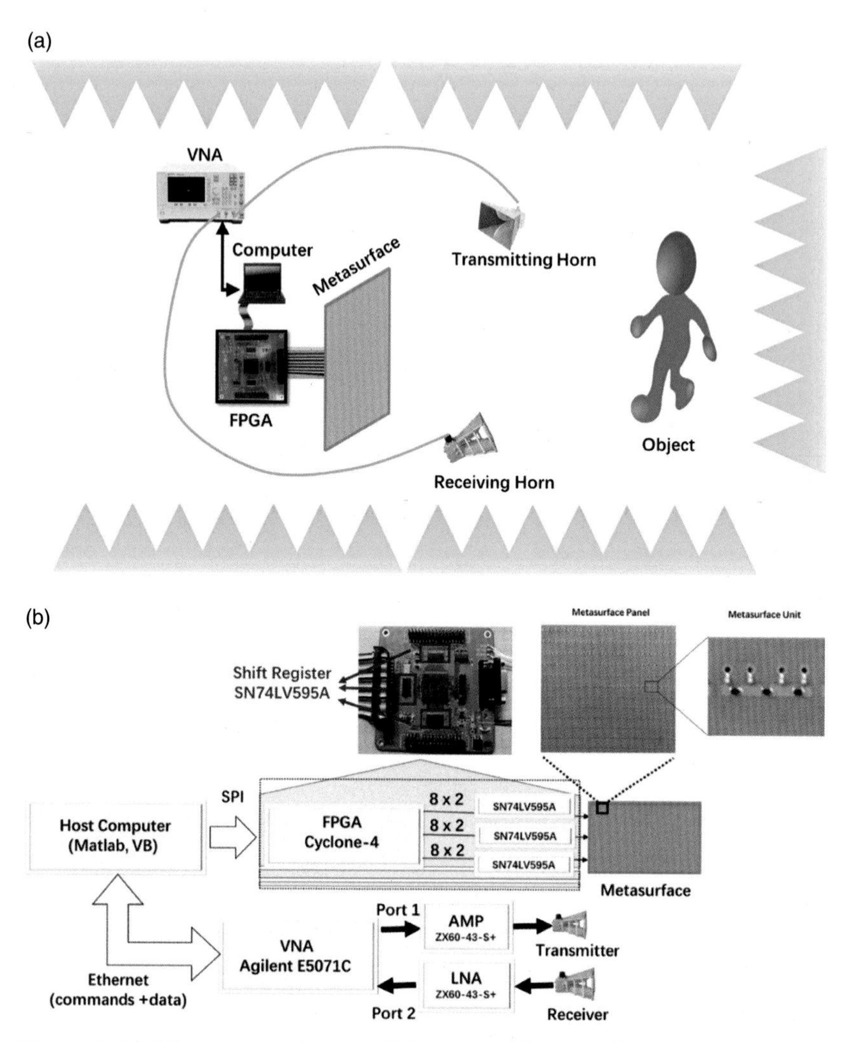

Figure 5.2 (a) The schematic map of the proposed proof-of-concept system. **(b)** Schematic block diagram of the proposed imager.[86]

under the control of FPGA generates the desired radiation patterns (i.e., the measurement modes), which are consistent with those required by machine learning. To that end, a straightforward two-step strategy was explored: the desirable radiation patterns are first trained by using the linear machine learning technique, and then the corresponding coding patterns of the metasurface are designed based on the obtained radiation patterns. As for the control patterns of the two-bit reprogrammable coding metasurface, the modified GS algorithm outlined in Section 2.4 is applied such that the desired radiation patterns can be produced. In particular, the four-phase quantization of the metasurface is performed in each iteration of forward/backward propagation. In addition, an effective induced current is introduced to characterize the realistic response of each meta-atom to address the accuracy issue arising from the point-like dipole model, which can be obtained by applying the Huygens principle or the source inversion technique.

5.3 Some Details about Algorithms

5.3.1 EM Modeling of Meta-atom and Metasurface

Here, we would like to provide some details about the EM modeling of the two-bit reprogrammable coding metasurface and meta-atom used in this section. As discussed previously, the technique of induced current inversion is explored to model the current source $J^{(S)}$ induced within the macro metasurface particles illuminated by a linearly polarized plane wave, where the superscript $S = 00$ (01, 10, 11) corresponds to the particle at the "00" ("01," "10," "11") state. For a metasurface hologram with $N_x \times N_y$ controllable macro meta-atoms illuminated by a plane wave, the resultant co-polarized radiation in the image plane is approximately written as

$$E(\boldsymbol{r}) = \sum_{n_x=1}^{N_x} \sum_{n_y=1}^{N_y} \int_{\Delta} g\left(\boldsymbol{r}, \boldsymbol{r}_{n_x,n_y} + \delta\boldsymbol{r}\right) J^{\left(S_{n_x,n_y}\right)}(\delta\boldsymbol{r}) d\delta\boldsymbol{r}, \tag{5.1}$$

where $g\left(\boldsymbol{r}, \boldsymbol{r}_{n_x,n_y}\right) = \frac{\exp(-jk_0|\boldsymbol{r}-\boldsymbol{r}_{n_x,n_y}|)}{4\pi|\boldsymbol{r}-\boldsymbol{r}_{n_x,n_y}|}$ is the three-dimensional free-space Green's function, k_0 is the wavenumber, and $\boldsymbol{r}_{n_x,n_y}$ denotes the central coordinate of the (n_x, n_y)th macro meta-atom. In Eq. (5.1), the integration is implemented over the surface of the macro meta-atom denoted by Δ, and the double summation is performed over all pixels of the metasurface. S_{n_x,n_y} represents the state of the (n_x, n_y)th pixel. Additionally, n_x and n_y denote the running indices of the pixel of the two-bit reprogrammable coding metasurface along the x and y directions, respectively.

The radiated electrical field **E** is linearly related to the current **J** induced over the metasurface by plane waves [92, 93]. The source inversion method is employed to calibrate the current distributions of $\mathbf{J}^{(00)}$, $\mathbf{J}^{(01)}$, $\mathbf{J}^{(10)}$, and $\mathbf{J}^{(11)}$. Then the macro meta-atom either at the state of "00," or "01," or "10," or "11" is illuminated by the plane wave, and the resulting co-polarized electric field is collected at one wavelength away from the macro meta-atom. The calibration measurements are organized into a column vector **e**. To characterize the inhomogeneous current induced on the macro meta-atom, the macro meta-atom is regarded as an array of 3×3 point-like dipoles, stacked in a nine-length column vector of **p**. Apparently, the calibration vector $\boldsymbol{e}$ is linearly related to the vector **p** through $\mathbf{e} = \mathbf{Gp}$, where the entries of **G** come from the three-dimensional Green's function in free space. Hence the complex-valued amplitude of **p** can be readily retrieved by solving the least squares problem of $\mathbf{e} = \mathbf{Gp}$.

5.3.2 Training Methods

The entries of the measurement matrix are approximately determined by $\boldsymbol{E}_i^T(\boldsymbol{r}_j) \cdot \boldsymbol{E}_i^R(\boldsymbol{r}_j)$, where $\boldsymbol{E}_i^R(\boldsymbol{r}_j)$ could be approximately regarded as that of a plane wave, since a single receiver is in the far-field region of scene. Then, the entries of the measurement matrix can be determined by the illumination pattern $\boldsymbol{E}_i^T(\boldsymbol{r}_j)$ of digital metasurface. In this way, the goal of machine learning amounts to find the desirable radiation patterns of reprogrammable coding metasurface. Taking the PCA training as an example, the operational procedure can be summarized as the following three-step method.

First, a set of training samples of optical images are collected, which can generate the theoretical PCA modes. Second, the modified G-S algorithm is implemented to produce the coding patterns of the digital metasurface in order to generate the desired radiation patterns whose amplitude distributions are consistent with those of theoretical PCA modes. Finally, in order to calibrate further possible errors of coding metasurface and to improve the quality of image reconstruction, the measurement matrix $\boldsymbol{H}$ is updated by solving the linear inverse problem of $\boldsymbol{Y} = \boldsymbol{HX}$, where the P-column matrix $\boldsymbol{X}$ denotes the collected P training optical images, and $\boldsymbol{Y}$ are the corresponding measurements. Note that this inverse problem can be solved by employing the standard least square method.

5.4 Results

5.4.1 Machine-Learning Guided Imaging

For the first demonstration, we consider the machine-learning metasurface imager to realize machine-learning-guided imaging. Throughout this section, two popular linear embedding techniques, namely the random projection and the principal component analysis (PCA), are applied to train the machine-learning imager. Apparently, it is trivial to conceive a random projection matrix by independently and randomly setting the status of PIN diodes of the reprogrammable metasurface. However, for PCA measurements, the status of PIN diodes needs to be carefully manipulated to achieve the desired measurement modes.

As the first set of investigations, the imaging performance of machine-learning image is tested over the MNIST dataset. For EM simulations, the objects are set to be perfect conduct in our numerical and experimental tests. We fabricated 5,000 training samples and 5,000 test samples. The theoretical PCA bases and corresponding experimental radiation patterns are shown in **Fig. 5.3a and b**, respectively, in which the control coding patterns of the metasurface according to the radiation patterns in **Fig. 5.3b** are presented in **Fig. 5.3c**. This set of figures illustrates that the two-bit reprogrammable coding metasurface can be well trained by the PCA. This means that it is capable of generating the measurement modes needed by the PCA, which can ensure the machine-learning-driven high-performance imaging from significantly reduced measurements. **Figure 5.3d–f** illustrates the recovered images of ten digit-like objects of the machine-learning imager trained with random projection and PCA with varying numbers of measurements. **Figure 5.3d** presents the retrieved images with ideally theoretical data from the random projection (the first row), PCA (the second row). In each method, the numbers of measurements are 10, 50, 100, 200, and 300 from the left to the right. We evaluated quantitatively the image qualities with respect to the signal-to-noise ratio (SNR) in **Fig. 5.3g**. Similarly, the imaging results from the full-wave simulation data are given in **Fig. 5.3e and h**, while the corresponding experimental results are demonstrated in **Fig. 5.3f and i**. It can be observed that the quality of image becomes better with the increasing measurements, and the PCA behaves better than the random projection, especially in the low measurement cases. Moreover, when the machine-learning metasurface imager is trained by the PCA, acceptable results can be achieved with a mere 100 measurements, corresponding to a compression rate of 12.8 percent.

As the second set of investigations, we experimentally examined the performance of the developed machine-learning imager by monitoring the movement of a person in front of the metasurface. In our study, we used one moving person to train our machine-learning imager, and used a different moving person to test it. The whole training time is less than 20 minutes with the proposed proof-of-concept system; however, such time could be remarkably reduced with a specialized receiver, as discussed in what follows. After being trained, the machine-learning imager can produce the measurement modes required by PCA. The 16 PCA leading measurement modes and the corresponding metasurface coding patterns are reported in **Fig. 5.4b and c**, respectively. Additionally, the corresponding theoretical PCA bases are reported in **Fig. 5.4a** as well. Here, the radiation patterns, at the distance of 1 m away from the metasurface, are experimentally obtained by performing the near-field scanning technique in combination with the near-to-far-field transform. **Figure 5.4a**–c demonstrates that the machine-learning imager is capable of generating the

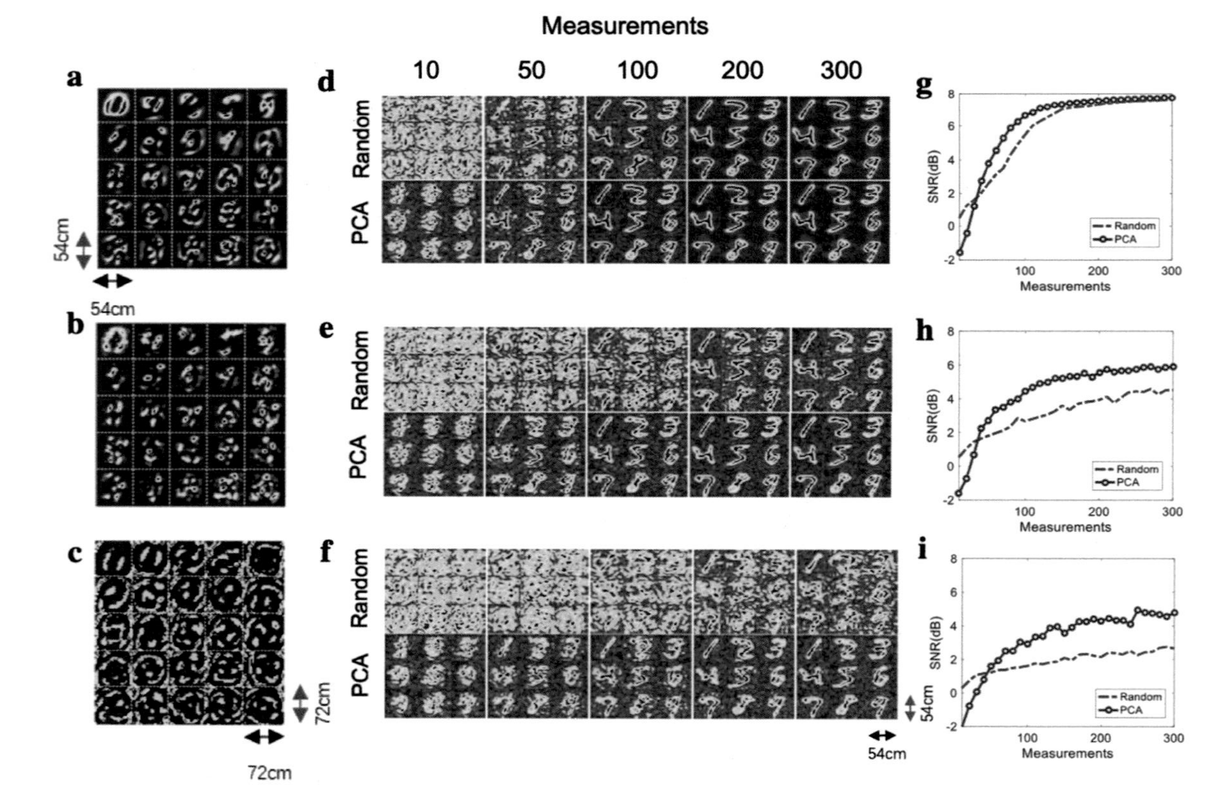

Figure 5.3 Machine-learning-driven imaging results by the real-time digital- metasurface imager trained with random projection and PCA. **(a)** The theoretical PCA bases. **(b)** The corresponding experimental radiation patterns. **(c)** The coding patterns of metasurface corresponding to the radiation patterns of **(b)**. **(d)** The theoretical results. **(e)** The full-wave simulation results. **(f)** The experimental results, in which the first and the second correspond to the random propagation and PCA, respectively. **(g–i)** The behaviors of image qualities (for **(g)** theory, **(h)** simulation, and **(i)** experiment results) characterized by SNRs as a function of the number of measurements.

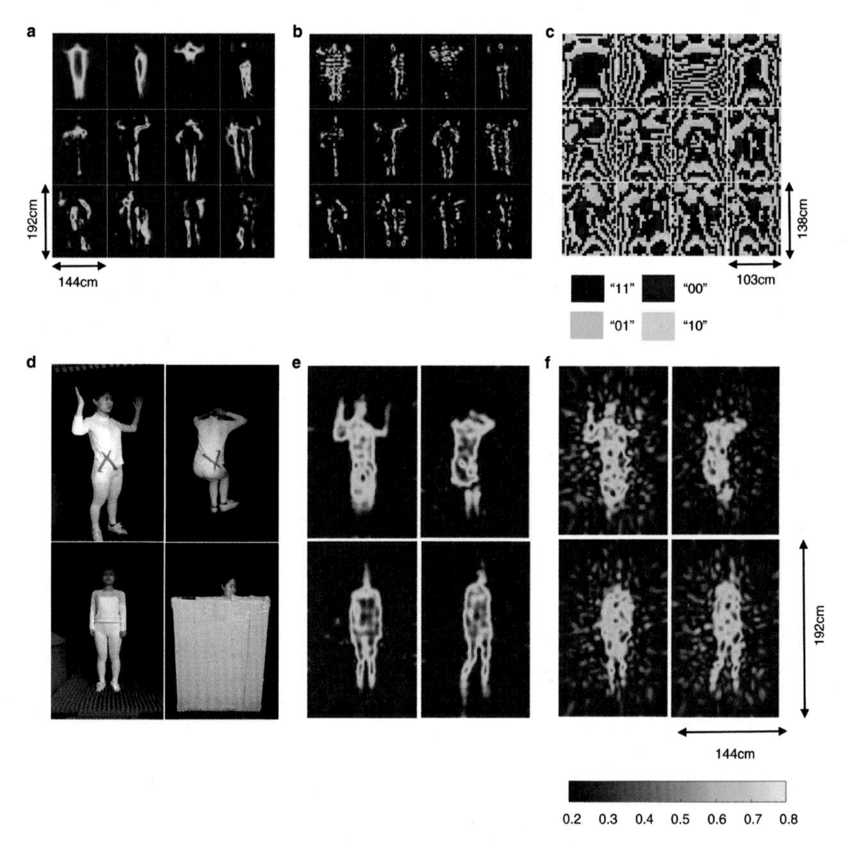

Figure 5.4 Configurations and results for human body imaging. **(a)** and **(b)** report 16 leading modes of theoretical PCA and the corresponding experimental patterns radiated by the machine-learning imager, respectively. **(c)** shows the coding patterns of the two-bit coding metasurface corresponding to **(b)**. **(d)** shows four images of a human body, where the top two people are armed with glass scissors, as shown in red. **(e)** and **(f)** are the reconstructed images corresponding to **(c)** using the machine-learning imagers trained with PCA and random projection, respectively, where 400 measurements are used.

measurement modes required by PCA, which establishes a solid foundation for the machine-learning-driven imaging with significantly reduced measurements.

Next, the trained machine-learning imager was used to monitor another moving person. In order to show the "see-through-the-wall" ability of the proposed machine-learning imager, selected reconstructions are plotted in **Fig. 5.4e**. In order to show the benefit gained by the machine-learning metasurface imager trained with PCA over the random projection. **Figure 5.4f** provides corresponding results when the reprogrammable metasurface is encoded in a random manner, that is, the measurement modes **H** behave as a random matrix. This set of figures clearly demonstrates that, since many relevant training samples are incorporated in PCA, the reconstructions have overwhelming advantages over those by the random projection in cases with a small amount of measurements. Note that both the head and feet cannot be clearly identified due to the limited FOV, which probably arises from two reasons, namely the use of a directive receiver (horn antenna), and the limited size of the metasurface with respect to the person under consideration. Nonetheless, it can be faithfully deduced from the previous results that the machine-learning imager, after being well trained, can be used to implement real-time and high-quality imaging from the considerably reduced measurements.

5.4.2 Object Recognition in Compressed Domain

In this section, we consider object recognition in the compressed domain using the machine-learning metasurface imager, which avoids the computational burden of costly image reconstructions.[92–93] Here, a simple nearest-neighbor algorithm [8, 10, 94], described in Section 5.1, is used to recognize the input image. In contrast to conventional machine learning techniques consisting of fully digital sampling followed by data postprocessing, our technique for object recognition is directly implemented at the physical level by using our reprogrammable metasurface.

Now, we consider the performance of the machine-learning metasurface imager on recognizing the body gestures, and corresponding experimental results are shown in **Fig. 5.5**. **Figure 5.5a** plots the classification performance of the machine-learning imager versus the number of measurements, as the imager is trained by the random projection (green dashed line) and PCA (red line with plus). For comparison, the theoretical PCA (blue straight line) result is also provided. Here, the classification rate is calculated by taking the average over three human-body gestures (standing, bending, and raising arms), in which each human-body gesture has over 1,000 test samples. **Figure 5.5b–d** compares specific classification results for the random projection, PCA, and theoretical PCA, where 25 measurements are used. Note that each period of data acquisition is about 0.016 seconds, and thus the whole data acquisition of 25 measurements is around 0.4 second. The classification performance of PCA is much better than the random projection, and it quickly approaches the ideal result with more measurements. In particular, 25 measurements are enough to get objection recognition of three human body actions, when the machine-learning imager is trained with PCA. This set of figures indicates that the machine-learning imager trained by PCA can achieve the

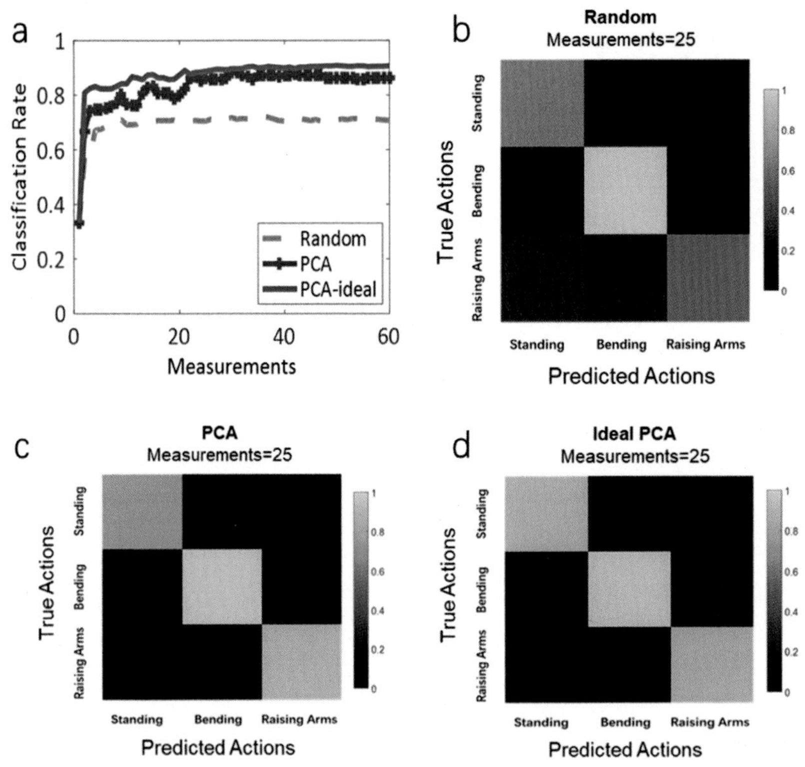

Figure 5.5 Classifying body gestures with the proposed imager. **(a)** Classification rate versus number of measurements when the programmable metasurface is trained by random projection and PCA. For comparison, the corresponding results by theoretical PCA (solid line) are also provided. **(b–d)** The comparison of specific classification results by **(b)** random projection, **(c)** PCA, and **(d)** theoretical PCA, in which 25 measurements are used.

theoretical upper limit of classification. Moreover, when the machine-learning metasurface imager is trained by the PCA, acceptable classification results can be achieved with around 60 measurements, corresponding to a compression rate of only 7.6 percent.

In this section, we discuss the machine-learning metasurface imager, showing that the imager can be utilized to remotely monitor and recognize the movement of human nearly without digital computation in a real-time manner. We can envision that such a sensing technique can be extended to other frequencies and beyond by using suitable tunable or switchable materials, for instance, thermal-sensitive phase-change materials (i.e., $Ge_2Sb_2Te_5$), gated graphene, and mechanical actuation. [6, 95]

6 Intelligent EM Metasurface Camera

Internet of things (IoT) and cyber physical systems have opened up many possibilities for smart societies, which are changing the way people live. In this era, there are ever-increasing demands in remotely monitoring people using radio-frequency probing signals. However, conventional systems can hardly be deployed in real-world settings, since they require target(s) to either deliberately cooperate or carry wireless active devices or identification tags. Recently, the intelligent EM metasurface camera has been proposed and demonstrated to be capable of monitoring the notable or nonnotable movements of multiple non-cooperative persons in a real-world setting, even when it is excited by the stray wireless signals existing everywhere rather than by the deliberately designed radio source [23]. In this section, we detail the intelligent EM metasurface camera in terms of the system design, algorithm, and performance. We expect that such a strategy could provide readers with new insights into the design of intelligent devices that are used to remotely sense and recognize more complicated human behaviors with negligible cost.

6.1 System Configuration

There is an ever-increasing demand to remotely probe where people are, what they are doing, what they want to express by their body language, and what their physiological states are, but in a way so as not to infringe on visual privacy. Up to this point, many RF sensing technologies have been proposed for various applications; for example, locating [96] and tracking [97–98], action recognition [99–100], human-pose estimation [101,102], breath monitoring [103,104], and so on. However, these systems are hindered from being deployed in real-world settings for the following reasons. First, they are typically designed for one specific task and rarely perform successive sensing missions adaptively, for instance, to instantly search for people of interest from a full scene and then adaptively recognize the subtle body features. Second, they require the target person to be deliberately cooperative and are challenging for monitoring non-cooperative targets; also, they involve weak signals that cannot be reliably distinguished from undesirable disturbances. Third, they suffer from complicated system designs and extremely expensive hardware cost, since a large number of sensors are needed to extract the fine-grained body information. In order to address these difficulties, Li et al. proposed the concept of an intelligent EM metasurface camera made by integrating reprogrammable metasurfaces with deep learning techniques – an inexpensive but intelligent sensing device, showing robust performance in sensing notable or nonnotable body signs [23]. The intelligent EM metasurface camera is conceptually shown in **Fig. 1a**, which relies on an Artificial Neural Network (ANN)-driven reprogrammable metasurface (called intelligent metasurface for short). Apparently, the large-aperture intelligent metasurface plays two critical roles: (1) manipulating the EM wave fields and thus realizing the smart data acquisition, and (2) processing the data flow in real time.

With reference to **Fig. 1a**, the intelligent metasurface camera has two operational modes: active and passive. In the active mode, the intelligent system is composed of a large-aperture reprogrammable metasurface, three CNNs, a transmitting antenna, a receiving antenna, and a vector network analyzer. In this setting, the camera includes a transmitter (Tx) to emit the RF signals into the investigation region through Antenna 1, and a receiver (Rx) to detect the echoes bounced from the subject through Antenna 2. In the passive mode, the camera has two or more coherent receivers to collect the stray wireless (here, Wi-Fi) waves bounced from the subject target. In this case, one antenna serves as a reference receiver to calibrate out the undesirable effects from the system error. An optical digital camera is synchronized with the whole intelligent metasurface, which is used to collect the labeled samples to train the deep ANNs.

The one-bit reprogrammable coding metasurface utilized here is composed of 32×24 electronically controllable meta-atoms with a size of 54×54 mm^2, and each meta-atom is integrated with a PIN diode. More details about the reprogrammable metasurface can be found in Section 2.2. The large-aperture reprogrammable coding metasurface is used to control the (ambient) EM wave field dynamically and adaptively by manipulating its control coding pattern through a FPGA, which has a two-fold role. Firstly, it serves as a relay station of information or an electronically controllable random mask, similar to the random mask in compressive sensing, transferring the EM signals carrying finer information of specimen to the receivers. Secondly, the reprogrammable metasurface with optimized coding patterns can focus the EM wave field toward the desired local spot(s) and meanwhile suppress the irrelevant inferences and clutters, which is extremely helpful in the recognition of body sign and vital sign, especially when the target is noncooperative.

Figure 6.1b illustrates schematically three building blocks of data flow pipeline, which includes three deep CNNs for data processing: IM-CNN-1, Faster R-CNN [105], and IM-CNN-2. IM-CNN-1 is an end-to-end mapping from the complex-valued microwave data to the desired images, as shown in **Fig. 6.2b**. IM-CNN-2 is a classifier to infer human hand signs from the microwave data, as shown in **Fig. 6.2c**, where the input is complex-valued microwave signal, and the output is the recognized label of hand signs. To improve the stability of the artificial neural network and meanwhile avoid

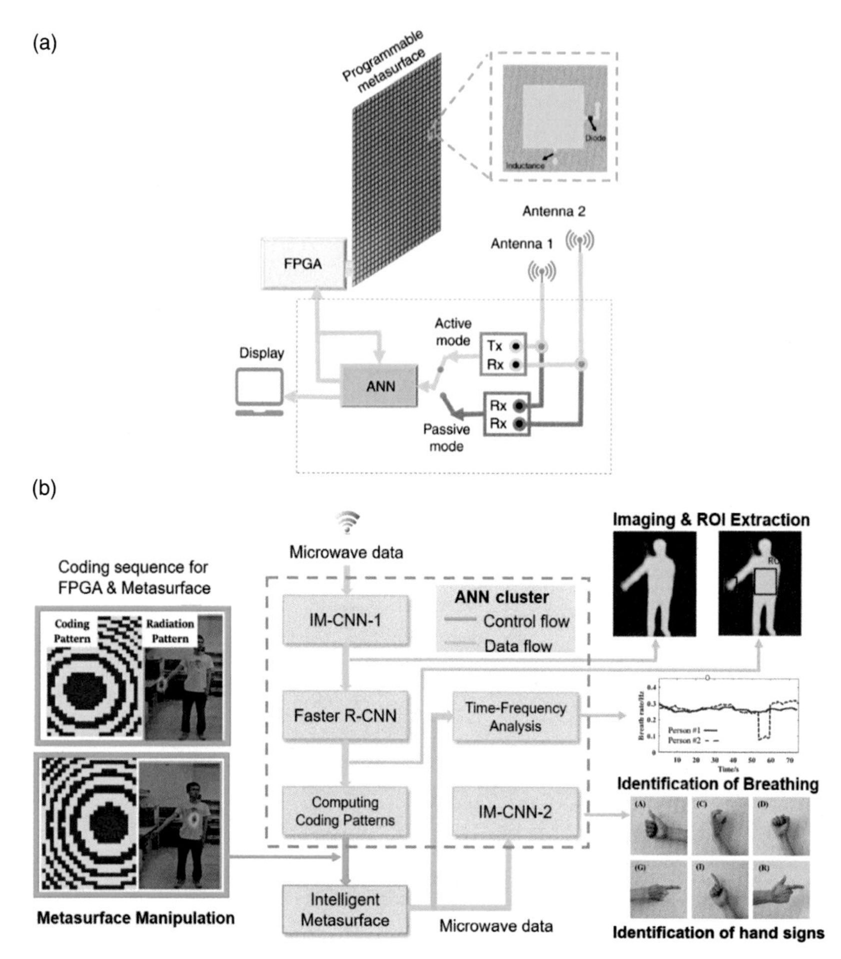

Figure 6.1 System configuration of intelligent metasurface camera. **(a)** The schematic configuration of an intelligent metasurface system by combining a large-aperture programmable metasurface for manipulating and sampling the EM wave fields adaptively with artificial neural networks (ANNs) for controlling and processing the data flow instantly. **(b)** The microwave data are processed with IM-CNN-1 to form the image of the whole human body.

the so-called gradient exploding and vanish, the proposed IM-CNN-1 and IM-CNN-2 are composed of a cascade of residual CNNs. The training procedure of IM-CNN-1 is briefly outlined in **Fig. 6.2a**. A commercial four-megapixel digital optical camera is embedded in the intelligent metasurface system to obtain a large number of labeled samples as calibrations for training IM-CNN-1. The labeled human-body images captured by the camera after background removal and binarization processing can be approximately regarded as the

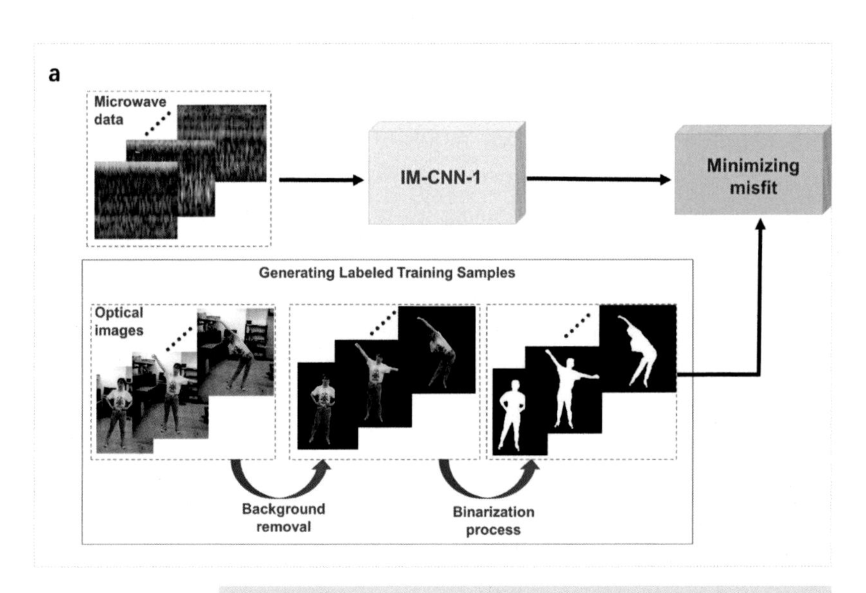

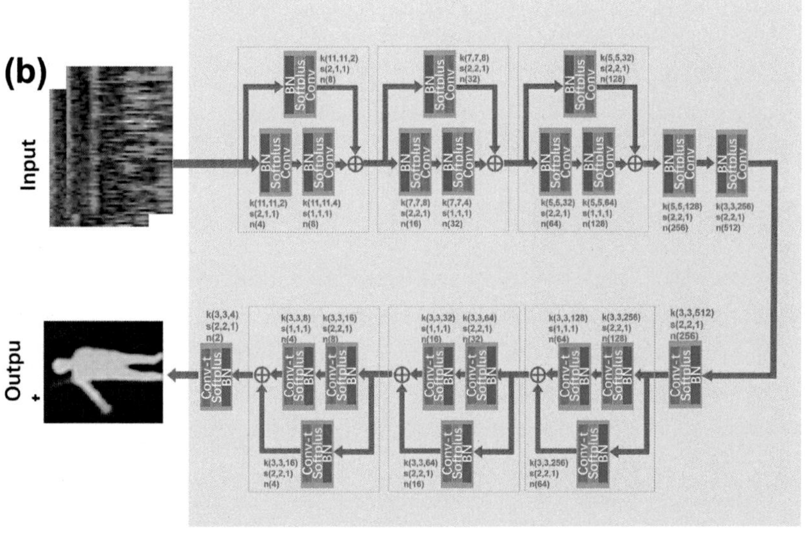

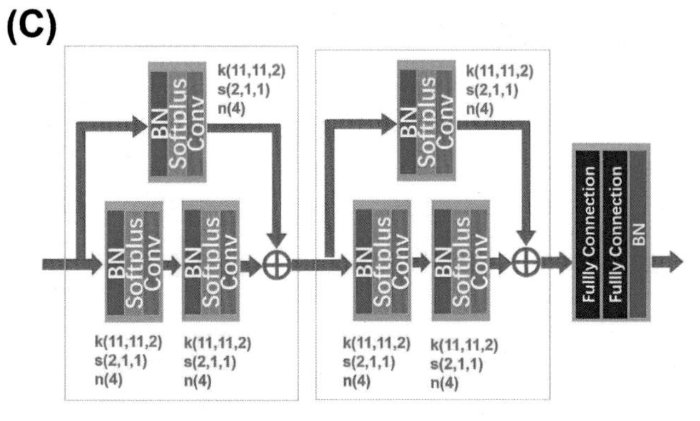

Figure 6.2 Designed artificial neural networks and their training scheme. **(a)** The supervised training procedure of IM-CNN-1. In our implementation, we

microwave reflectivity images of the human body, because the microwave reflection of the human body can be approximated to be homogenous while the body is undergoing frequencies from 2.4 to 2.5 GHz.

The training stage is done using the Adam optimization method, with mini-batches size of 32, and epoch setting as 50. The learning rates are set to 10^{-4} and 10^{-5} for the first two layers and the last layer in each network, respectively, and halved once the error plateaus. The complex-valued weights and biases are initialized by random weights with zero-mean Gaussian distribution and standard deviation of 10^{-3}. The computations are performed with an AMD Ryzen Threadripper 1950X 16-Core processor, NVIDIA GeForce GTX 1080Ti, and 128 GB access memory. The networks are designed using the Tensor Flow library.

6.2 Signal Model

Here, we discuss the signal model of the intelligent metasurface camera. To this end, we assume that the subject target characterized by its reflectivity $O(r')$ is situated in a real-world indoor environment, and consider the aforementioned active and passive sensing modes.

6.2.1 Signal Model for Active Sensing Mode

In this setting, the metasurface camera system has a transmitter (Tx) for actively emitting the EM waves into the investigation region through Antenna 1, and a receiver (Rx) for receiving the echoes bounced off the subject specimen through Antenna 2, as shown in **Fig. 6.1a**. Additionally, the subject is at a distance of 0.5 m–1.0 m away from the front side of a one-bit reprogrammable

---

Caption for Figure 6.2 (cont.)

use a four-megapixel digital optical camera to achieve the required labeled training samples. Considering that the reflectivity of the human body is approximated to be uniform over the whole body, undergoing frequencies from 2.4 GHz to 2.5 GHz, the color optical images are processed with a binarization process as our labeled training samples. **(b)–(c)** are the architectures of proposed IM-CNN-1 and IM-CNN-2, in which BN denotes the batch normalization, softmax denotes the soft-max nonlinear activation function, relu is the ReLu activation function, $\mathrm{k(a, b, c)}$ denotes the convolutional kernel with a size of $\mathrm{a \times b \times c}$, and $\mathrm{n(a)}$ denotes the number of convolutional kernels as a, and CNN-t denotes the transpose operation of CNN.

coding metasurface. Assuming that the Tx antenna at $\boldsymbol{r}_t$ emits periodically the RF signal of $s(\omega)$, where ω is angular frequency. With such illumination, the programmable metasurface configured with the mth coding pattern will give rise to the secondary radiation denoted by $u_m(\boldsymbol{r}';\omega)s(\omega)$ ($\boldsymbol{r}' \in \Omega$) inside the investigation region Ω. Under the Born assumption, the electrical field acquired by the receiver at $\boldsymbol{r}_R$ reads:

$$E_m(\boldsymbol{r}_R;\omega) = s(\omega)\int_{\Omega} G(\boldsymbol{r}_R,\boldsymbol{r}';\omega)u_m(\boldsymbol{r}';\omega)O(\boldsymbol{r}')d\boldsymbol{r}', \quad m = 1, 2, \ldots, M. \tag{6.1}$$

For simplicity, we have restricted ourselves to the scalar case. However, the analysis can be readily extended to the full-vector case in a straightforward manner. Here, M denotes the total number of coding patterns of metasurface, $G(\boldsymbol{r}_R,\boldsymbol{r}';\omega)$ denotes the Green's function of indoor environment, and $u_m(\boldsymbol{r}';\omega)$ depends on the coding patterns of the metasurface. Now, the reflectivity function $O(\boldsymbol{r}')$ can be retrieved from the measurements $\{E_m(\boldsymbol{r}_R;\omega), m = 1, 2, \ldots, M\}$ by solving **Eq. (6.1)**. However, it is not a trivial issue because, for the complicated indoor environment, it remains a challenging open topic to characterize and analyze $G(\boldsymbol{r}_R,\boldsymbol{r}';\omega)$ and $u_m(\boldsymbol{r}';\omega)$ in a tractable manner. To address this problem, deep learning is explored. Specifically, we developed a specialized end-to-end deep learning network, for example IM-CNN-1, to solve Eq. (6.1), which establishes a nearly closed-form solution mapping from the measurements $\{E_m(\boldsymbol{r}_R;\omega), m = 1, 2, \ldots, M\}$ to the desirable image of $O(\boldsymbol{r}')$.

6.2.2 Signal Model for Passive Sensing Mode

In this setting, the intelligent metasurface illuminated by noncooperative EM sources will give rise to the secondary EM radiation, and such radiation will further illuminate the subject target. The intelligent metasurface in the passive mode has two (or more) coherent receivers, which are connected to two ports of an oscilloscope (Agilent™ MSO9404A) for the data acqusition.

When illuminated by a noncooperative signal $s_m(\omega)$, the intelligent metasurface configured with the mth coding pattern within a time interval will give rise to the secondary radiation denoted by $u_m(\boldsymbol{r}';\omega)s_m(\omega)(\boldsymbol{r}' \in \Omega)$ inside the investigation domain Ω. Herein, the subscript m of $s_m(\omega)$ highlights that the noncooperative Wi-Fi signals varies with the change of coding sequence of the metasurface in an *unknown* way. It is applicable when the data frames of commodity Wi-Fi signals are considered. By contrast, if we only consider the beacon frames of commodity Wi-Fi signals, the unknown signal $s_m(\omega)$ is well defined and is nearly unchanged with respect to m. Under the Born approximation, the electrical fields at $\boldsymbol{r}_1$ and $\boldsymbol{r}_2$ scattered from the subject targets are expressed as

$$E_m(\boldsymbol{r}_1;\omega) = s_m(\omega)\int_{\Omega} G_m(\boldsymbol{r}_1,\boldsymbol{r}';\omega)u_m(\boldsymbol{r}';\omega)O(\boldsymbol{r}')d\boldsymbol{r}' \tag{6.2}$$

and

$$E_m(\boldsymbol{r}_2;\omega) = s_m(\omega)\int_{\Omega} G_m(\boldsymbol{r}_2,\boldsymbol{r}';\omega)u_m(\boldsymbol{r}';\omega)O(\boldsymbol{r}')d\boldsymbol{r}' \tag{6.3}$$

in which $m = 1,2,\ldots,\ M$; $G_m(\cdot,\cdot:\omega)$ is the Green's function in the indoor environment including the metasurface, and the subscript m implies its dependence on the coding pattern of metasurface. In order to calibrate out the unknown signal $s_m(\omega)$, we take the correlation average between $E_m(\boldsymbol{r}_1;\omega)$ and $E_m(\boldsymbol{r}_2;\omega)$, in particular,

$$\begin{aligned}\langle E_m(\boldsymbol{r}_1;\omega)E_m^*(\boldsymbol{r}_2;\omega)\rangle &= \langle|s_m(\omega)|^2\rangle\int_{\Omega}\int_{\Omega} d\boldsymbol{r}'d\boldsymbol{r}''\, G_m(\boldsymbol{r}_1,\boldsymbol{r}';\omega)\\ &G_m^*(\boldsymbol{r}_2,\boldsymbol{r}';\omega)u_m(\boldsymbol{r}';\omega)u_m^*(\boldsymbol{r}'';\omega)O(\boldsymbol{r}')O^*(\boldsymbol{r}'')\end{aligned} \tag{6.4}$$

where $\langle\cdot\rangle$ denotes the ensemble average operation. We assume that the noncooperative Wi-Fi signals and associated radiations are statistically stationary. Thus $\langle|s_m(\omega)|^2\rangle$ is almost unchanged with respect to m.

Taking the preceding observations into account, we perform the following steps to achieve the estimation of $\langle E_m(\boldsymbol{r}_1;\omega)E_m^*(\boldsymbol{r}_2;\omega)\rangle$ for each coding pattern:

Step 1. Acquire a 50 μs-long observation signal for each receiver, and consequently, $y(t,\boldsymbol{r}_1)$ and $y(t,\boldsymbol{r}_2)$, for receivers 1 and 2, respectively.

Step 2. Divide $y(t,\boldsymbol{r}_1)$ and $y(t,\boldsymbol{r}_2)$ randomly into 200 overlapped 3 μs-long signal samples $\{y^{(i)}(t,\boldsymbol{r}_1),\ i = 1,2,\ldots,200\}$ and $\{y^{(i)}(t,\boldsymbol{r}_2),\ i = 1,2,\ldots,200\}$.

Step 3. Perform FFTs on above signal samples, that is, $FFT\{y^{(i)}(t,\boldsymbol{r}_1)\},\ i = 1,2,..,200$ and $FFT\{y^{(i)}(t,\boldsymbol{r}_2)\},\ i = 1,2\ldots200$.

Step 4. Calculate the ensemble average

$$\langle E_m(\boldsymbol{r}_1;\omega)E_m^*(\boldsymbol{r}_2;\omega)\rangle = \frac{1}{200}\sum_{i=1}^{200} FFT\left\{y^{(i)}(t,\boldsymbol{r}_2)\right\}\left[FFT\left\{y^{(i)}(t,\boldsymbol{r}_2)\right\}\right]^*.$$

If we consider the indoor environment except for the intelligent metasurface as a random process, then Eq. (6.4) can be expressed as

$$\langle E_m(\boldsymbol{r}_1;\omega)E_m^*(\boldsymbol{r}_2;\omega)\rangle = \langle|s_m(\omega)|^2\rangle\int_\Omega\int_\Omega d\boldsymbol{r}'d\boldsymbol{r}''\, G_m(\boldsymbol{r}_1,\boldsymbol{r}';\omega)G_m^*(\boldsymbol{r}_2,\boldsymbol{r}';\omega)u_m(\boldsymbol{r}';\omega)u_m^*(\boldsymbol{r}'';\omega)O(\boldsymbol{r}')O^*(\boldsymbol{r}'')$$
$$m = 1,2,\ldots,M. \tag{6.5}$$

For simplicity, we take numerical discretion of the investigation domain Ω, so that the following approximations hold:

$$\langle u_m(\boldsymbol{r}';\omega)u_m^*(\boldsymbol{r}'';\omega)\rangle \approx \langle|u_m(\boldsymbol{r}';\omega)|^2\rangle\delta(\boldsymbol{r}-\boldsymbol{r}') \tag{6.6}$$

$$\text{and/or } \langle G_m(\boldsymbol{r}_1,\boldsymbol{r}';\omega)G_m^*(\boldsymbol{r}_2,\boldsymbol{r}';\omega)\rangle \approx \langle|G_m(\boldsymbol{r}_1,\boldsymbol{r}';\omega)|^2\rangle\delta(\boldsymbol{r}-\boldsymbol{r}'). \tag{6.7}$$

Note that Eq. (6.7) can be justified in terms of the time-reversal theory in random surrounding media or cavity. As a consequence, Eq. (6.5) becomes

$$\langle E_m(\boldsymbol{r}_1;\omega)E_m^*(\boldsymbol{r}_2;\omega)\rangle = \langle|s_m(\omega)|^2\rangle\int_\Omega d\boldsymbol{r}'\langle|G_m(\boldsymbol{r}_1,\boldsymbol{r}';\omega)|^2\rangle\langle|u_m(\boldsymbol{r}';\omega)|^2|\mathrm{O}(\boldsymbol{r}')|^2\rangle. \tag{6.8}$$

It can be observed (Eq. (6.5) or Eq. (6.8)) that the coherent measurements between a pair of receivers $\{\langle E_m(\boldsymbol{r}_1;\omega)E_m^*(\boldsymbol{r}_2;\omega)\rangle,\ m = 1,2,\ldots,M)$ behave as a function of the reflectivity of the subject target $\mathrm{O}(\boldsymbol{r}')$. Although the relation cannot be analyzed or tackled in an analytical way, the deep learning technique can be utilized to retrieve $\mathrm{O}(\boldsymbol{r}')$ from coherent measurements $\{\langle E_m(\boldsymbol{r}_1;\omega)E_m^*(\boldsymbol{r}_2;\omega)\rangle,\ m = 1,2,\ldots,M\}$.

6.3 Performance Evaluation

6.3.1 In Situ High-Resolution Imaging

We present the in situ high-resolution microwave imaging of the whole human body in the active mode, which is conducted in our lab environment. The *kernel* of the intelligent metasurface for whole-body imaging is IM-CNN-1 to process the microwave data instantly. To obtain a large number of labeled samples for training IM-CNN-1, a commercial four-megapixel digital optical camera is embedded in the metasurface system. The training samples captured by the optical camera are used to train IM-CNN-1 after being preprocessed with background removal, threshold saturation, and binary-value processing. The labeled human-body images can be approximately regarded as EM reflection images of the human body over the undergoing frequencies from 2.4 to 2.5 GHz.

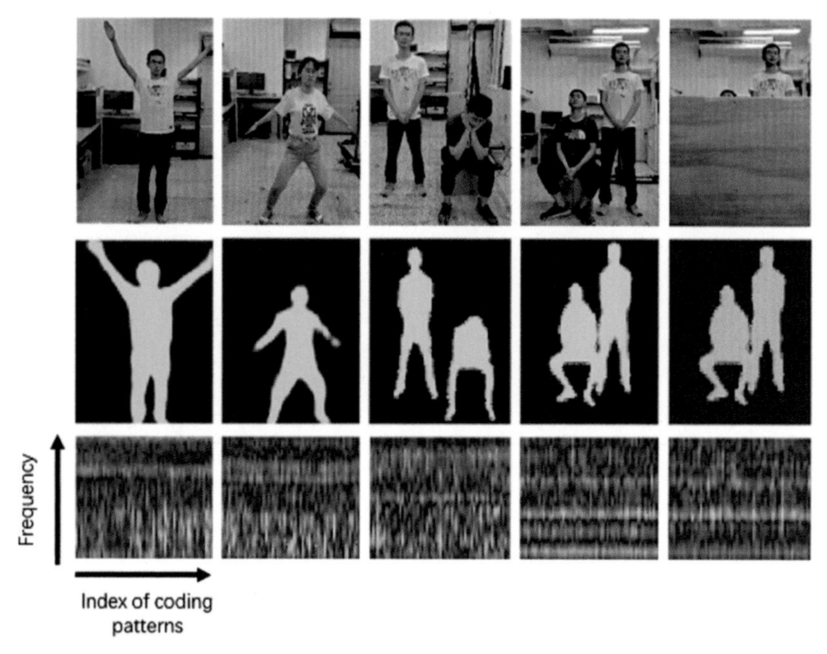

Figure 6.3 In-situ imaging results using the intelligent metasurface with active microwave. **(a)** The first row shows the optical images of specimen, which include single person with different gestures, two persons with different gestures, and two persons behind a 5-cm-thick wooden wall. **(b)** The second row illustrates the corresponding imaging results by the intelligent metasurface with IM-CNN-1. **(c)** The bottom row presents corresponding amplitudes of microwave data.

We experimentally characterize the performance of the metasurface system in achieving high-resolution images of the whole human body and simultaneously monitoring notable movements in an indoor environment. The trained metasurface is used to produce high-resolution images of the test persons, from which their body gesture information can be identified. A set of results is presented in **Fig. 6.3.** Particularly, the "see-through-the-wall" ability is validated by clearly detecting notable movements of the test persons behind a 5-cm-thick wooden wall. Selected results are provided in the rightmost column of **Fig. 6.3**, where the corresponding optical images and microwave raw data are provided as well.

To examine the imaging quality quantitatively, **Figure 6.4** compares the image quality versus the number of random coding patterns of the programmable metasurface in terms of the similarity structure index metric (SSIM). We show that it is enough to achieve high-quality images by using

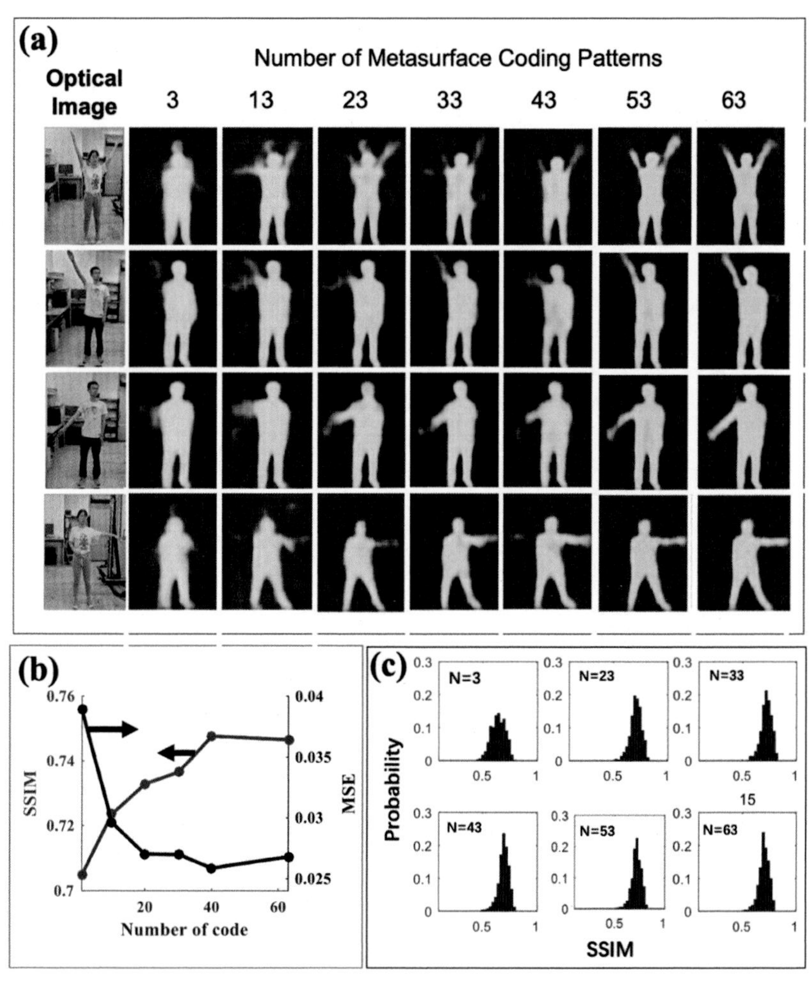

Figure 6.4 The image quality of in situ high-resolution imaging with the different numbers of coding patterns of programmable metasurface. **(a)** Selected samples of images for different numbers of coding patterns: 3, 13, 23, 33, 43, 53, and 63. **(b)** The SSIMs and SMEs as functions of the number of coding patterns. **(c)** The statistical analysis of image quality in terms of SSIM.

53 coding patterns, where 101 frequency points from 2.4 to 2.5 GHz are utilized for each coding pattern. Note that the switch time of coding patterns is around 10 μs, implying that the time in data acquisition is less than 0.7 ms in total even if 63 coding patterns are used. Then, we conclude that the intelligent metasurface integrated with IM-CNN-1 can instantly produce high-quality images of multiple persons in the real world, even when they are behind obstacles.

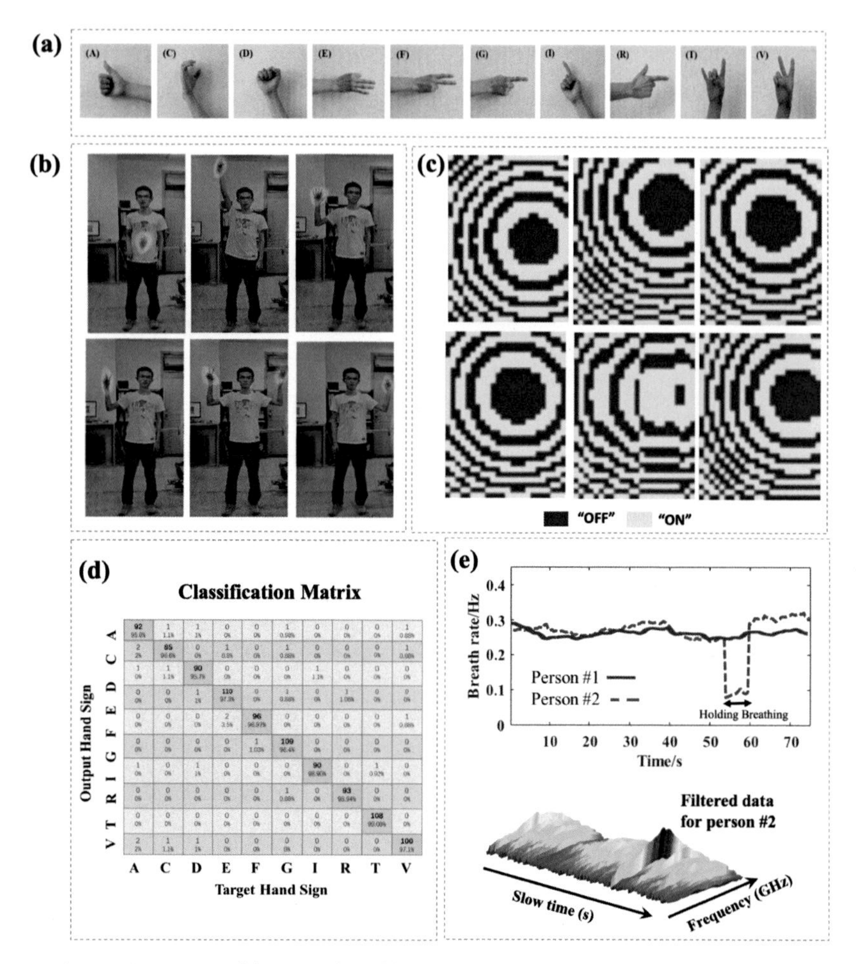

Figure 6.5 Recognition results of human hand signs and respirations by the intelligent metasurface with active microwave. **(a)** Ten hand signs of English letters considered in this work. **(b)–(c)** Selected results of the microwave radiations focused on the desirable spots, for instance, human hands and chest, and corresponding optimized coding patterns of programmable metasurface. **(d)** The classification matrix of 10 hand signs in (a) obtained by using the IM-CNN-2. **(e)** Results of human respiration of two persons in our lab environment, where person #1 has normal breathing, and person #2 holds his breath for around 55 s.

6.3.2 Instant Recognition of Hand Signs and Vital Signs

After obtaining the high-resolution image of the whole body, the intelligent metasurface is then used to recognize the hand signs and vital signs adaptively in real indoor environments. This capacity benefits from the robust feature of the intelligent metasurface in adaptively focusing the EM energy to the desired spots with very high spatial resolution. This feature supports accurate detections of EM echoes reflected from the human hand for recognizing the sign language or from the chest for identifying respiration. Typically, the rate of the human hand sign language and respiration is in the order of 10–30 bps, which is drastically slower than the switching speed of coding patterns by a factor of 10^5. Thus, the radiation beams of the intelligent metasurface are manipulated to rapidly scan the local body parts of interest in each observation time interval. We realize monitoring the hand signs and respirations of multiple people simultaneously in a time-division multiplexing way.

To fulfil this complicated task, we propose a three-step routine procedure. Firstly, the Faster R-CNN[47] is applied to extract the hand or chest part from the full-scene image obtained with IM-CNN-1 in a divide-and-conquer manner. Secondly, the metasurface is manipulated by adaptively changing its coding pattern to make its radiation beam point to the hand or chest (see **Fig. 6.5a–c**). Thirdly, IM-CNN-2, an end-to-end mapping from the microwave data to the label of hand-sign language, is developed to recognize the hand signs. The conventional time-frequency analysis is performed for detecting the respiration.

The training samples of IM-CNN-2 include ten hand signs and 8000 samples for each hand sign. **Figure 6.5d** reports the classification matrix for the ten hand signs with average recognition accuracy of above 95 percent by the metasurface system integrated with IM-CNN-2, where the test people are behind a 5-cm-thickness wooden wall. We clearly see that the performance of hand-sign recognition is almost not affected by the number of test persons after the hand parts are well identified by Faster-R-CNN. The respiration is an important health metric for tracking human physiological states (e.g., sleep, pulmonology, and cardiology). Similar to the recognition of human hand signs, we use the intelligent metasurface to monitor human respiration with high accuracy. **Figure 6.5e** reports the monitored results of the respiration of two test persons behind a wooden wall. We observe that the normal breathing and holding of the breath are clearly distinguished, and the respiration rate can further be identified with the accuracy of 95 percent and beyond, where the ground truth is achieved by a commercial breathing monitoring device. It can be expected that the identification performance is almost independent of the number of test persons due to the use of time-division multiplexing respiration detection.

6.3.3 Imaging and Recognition with Passive Modes

The intelligent metasurface system works at around 2.4–2.5 GHz, which is exactly the frequency of commodity Wi-Fi signals [106–108]. Here, we investigate the performance of high-resolution imaging of full scene and recognition of human hand signs and vital signs when the metasurface is *excited by the commodity stray Wi-Fi signals*. For simplicity, we consider the use of Wi-Fi beacon signals. In this case, the metasurface system works in three major aspects: (1) the stray Wi-Fi signals are dynamically manipulated by the metasurface; (2) two or more coherent receiving antennas are used to acquire the Wi-Fi signals bounced from the subject with the aid of an oscilloscope (Agilent^{TM} MSO9404A); (3) the microwave data acquired by receivers are coherently preprocessed before sending them to IM-CNN-1 such that the uncertainties on stray Wi-Fi signals can be calibrated out.

Figure 6.6a presents a set of in situ passive imaging results of the subject person behind the wooden wall in our indoor lab environment, where random coding patterns are also used in the programmable metasurface. We note that the imaging results by the commodity stray Wi-Fi signals are comparable to those in the active mode. Based on the high-resolution images of the full human body, we can realize the recognition of hand signs and vital signs by adaptively performing the three-step routine procedure in the active mode. In particular, the Faster R-CNN is operated on the full-scene image to instantly find the location of a hand or the chest; then suitable coding patterns of the intelligent metasurface can be achieved and controlled so that the stray Wi-Fi signals are spatially focused and enhanced on the desired spots; and finally, IM-CNN-2 or the time-frequency analysis algorithm is used to realize the recognition of hand signs and vital signs. As shown in **Fig. 6.6b–c**, the commodity Wi-Fi signals can be well focused on the desired location, for example the left hand of the subject person, by using the developed intelligent metasurface. As a result, SNR of Wi-Fi signals can be significantly enhanced with a factor of more than 20 dB, which is directly helpful in the subsequent recognitions of hand signs and vital signs. **Figure 6.6d and e** show experimental results of the hand-sign and respiration recognitions of two people, which have better accuracies of 90 percent and 92 percent, respectively. To summarize, even with the illumination of stray Wi-Fi signals, the proposed intelligent metasurface can realize high-resolution images of full-scene and high-accuracy recognitions of hand signs and vital signs of multiple people in a smart and real-time way in the real world.

To summarize, we discuss the intelligent metasurface camera in terms of system design, signal model, and sensing performance, showing its robust performance in remotely monitoring notable human movements, subtle gesture languages, and physiological states from multiple noncooperative people in real-world settings. We show that such a camera works well under passive stray Wi-Fi signals, in which the one-bit reprogrammable coding metasurface supports adaptive manipulations and smart acquisitions of the stray Wi-Fi signals. It introduces a new way to not only "see" what people are doing, but also "hear"

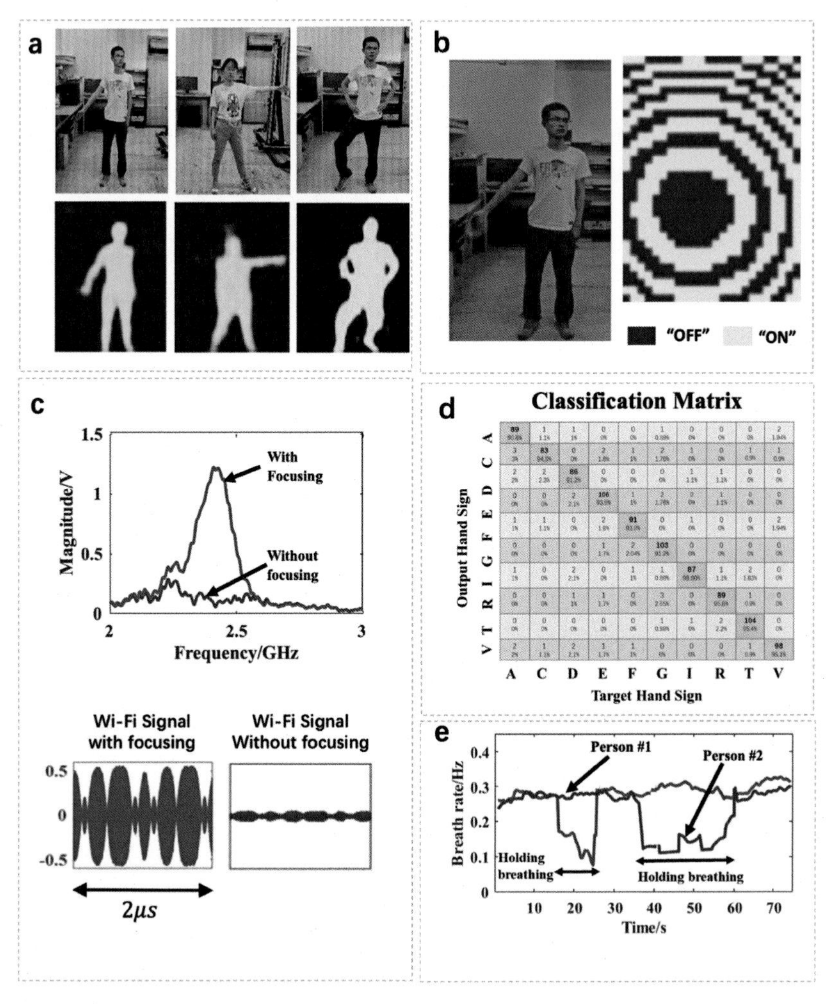

Figure 6.6 Experimental results of in situ imaging, hand-sign recognition, and respiration identification using the intelligent metasurface in the passive mode with commodity stray Wi-Fi signals. **(a)** In situ imaging results using the intelligent metasurface excited with commodity Wi-Fi signals. The first row shows the optical images of the subject person with different gestures behind a 5-cm-thick wooden wall. **(b)** On the left is the result of the Wi-Fi signals focused on the desired spot of a human hand, and on the right is corresponding coding pattern of the programmable metasurface. **(c)** The Wi-Fi signals with and without being focused through the programmable metasurface have been compared. The top row compares the frequency spectrums of Wi-Fi signals. **(d)** The classification matrix of 10 hand signs obtained by using the IM-CNN-2. **(e)** Results of human respiration of two noncooperative persons behind a 5-cm-thick wall in our lab environment.

what people talk about without deploying any acoustic sensors, even when multiple people are behind obstacles. We expect that the concept of intelligent metasurface can be extended over the entire EM spectrum, and could open a new avenue for future smart homes, human-device interactive interfaces, health monitoring, and safety screening.

7 Intelligent Integrated Metasurface Sensors

An entire sensing pipeline consists of three building blocks: the scene, data acquisition, and digital data processing; and an ideal intelligent sensing system needs to consider them as a whole. In this sense, the sensing strategies discussed in previous sections lack "intelligence": they indiscriminately acquire all information, ignoring available knowledge about the scene, sensing task, and hardware constraints. Yet, using this available a priori knowledge is critical to the data acquisition with *relevant* information – the crucial conceptual improvement necessary to reduce latency and computational burden. To fully reap the benefits of artificial intelligence techniques in EM sensing, the scene, acquisition, and processing need to jointly learned in a unique pipeline, called intelligent integrated sensing (IIS) for short. Here, we discuss two recent advances in this emerging direction: the IIS based on the variational auto-encoder (VAE) [24], and the IIS based on the free-energy minimization [109]. For both of them, the settings of data acquisition and processing are jointly optimized for the given hardware, task, and expected scene. We expect that such integrated sensing methodologies could pave the way to low-latency sensing for human–computer interaction, health care, and security screening.

7.1 VAE-Based Integrated Sensing

Cost efficiency and speed remain two major challenges for the deployment of EM sensors in many practical applications, especially for the purpose of real-time in situ sensing. From previous discussions, we know that most novel hardware solutions share the spirit of compressive sensing, that is, the information to be acquired is multiplexed across independent frequencies or spatial channels, and the pressure on the hardware cost is shifted toward the software level. In Section 6, data processing on the digital level is highly efficient through the utilization of machine learning techniques and powerful computation resources, and is usually thought to be intelligent because the information of the scene has been considered through a large amount of training samples. However, it still lacks the intelligence, since it adopts the generic (random) scene illumination, ignoring available knowledge about scene, sensing task, and hardware constraints. Yet, using this available a priori knowledge is critical to data acquisition with *relevant* information – the crucial conceptual improvement necessary to reduce latency and computational burden. A first step to add intelligence consisted in adapting the scene illuminations to the knowledge of what scene is expected via a principal component analysis (PCA) of the expected scene [16]. Albeit yielding significant performance improvements, this technique still considers acquisition and processing separately, and hence fails to use task-specific measurement settings that highlight salient features for the processing layer. To address these difficulties, the sensing system needs to simultaneously optimize the sets of scene, acquisition, and processing in a unique pipeline, in a process called intelligent integrated sensing. Recently, we have proposed intelligent integrated sensing by exploring the reprogrammable coding metasurface for data-driven learnable data acquisition and integrating it into a data-driven learnable data processing pipeline. Thereby, the measurement strategy can be learned jointly with a matching data postprocessing scheme, optimally tailored to the specific sensing hardware, task, and scene.

We note that there are multiple related works in optics [110–114] and a similar concept has recently also been studied in the context of ultrasonic imaging [115]. While current EM sensing systems are designed to first generate high-resolution images as a checkpoint for subsequent AI-aided recognition tasks, the resulting acquisition of irrelevant information and the huge flux of data to be processed cause significant inefficiencies. Hence, for tasks like object recognition it is most efficient to skip the intermediate imaging step and to directly process the raw data.

7.1.1 System Design

The intelligent sensing system, as shown in **Fig. 7.1**, consists of a pair of horn antennas, a vector network analyzer, and a large-aperture programmable metasurface. The operational principle of the presented sensing system is described as follows. Antenna 1, connected to port-1 of VNA, is used to emit microwave illumination signals, which are spatially shaped by the m-ANN-driven reprogrammable metasurface. After being scattered by the subject of interest, the wave field shaped by the metasurface is received by Antenna-2 connected to port-2 of VNA. Finally, the received raw data are instantly processed by the r-ANN, producing the desired imaging or recognition results. The one-bit reprogrammable coding metasurface consists of 32×24 meta-atoms operating at around 2.4 GHz. In our implementation, the switching time of the coding pattern of the metasurface was set to 20 μs, thus the time of data acquisition was on the order of $M \times 20$ μs, where M is the total number of the coding patterns. More details about the reprogrammable metasurface considered here can be found in Section 2.2.

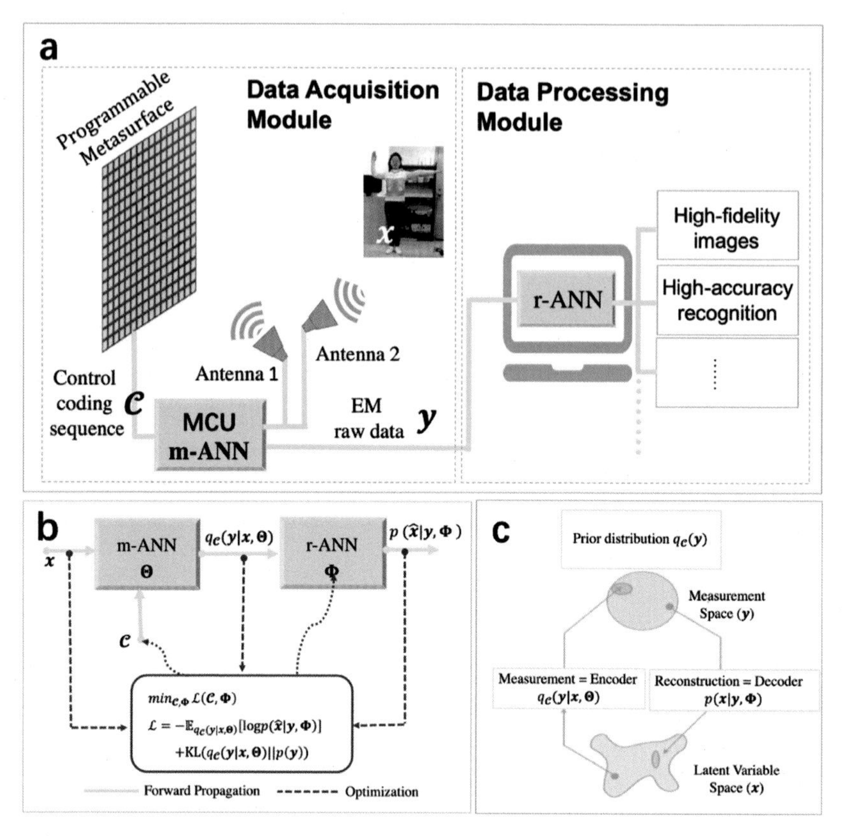

Figure 7.1 Setup and working principle of the learned EM sensing system. **(a)** The intelligent sensing system consists of two data-driven learnable modules: the *m-ANN*-driven data acquisition module and the *r-ANN*-driven data processing module. **(b)** The *m-ANN*-based model of data acquisition as a trainable physical network is fully integrated with the *r-ANN*-driven data processing pipeline into a unique sensing chain. **(c)** Our interpretation of the entire sensing process in the VAE framework is as follows: the latent variable space, $\boldsymbol{x}$, is encoded by the analog measurements in a measurement space, $\boldsymbol{y}$; the digital reconstruction decodes the measurements to return to the latent variable space.

The intelligent hardware layer relies on a reprogrammable metasurface to shape the waves illuminating the scene. We introduced a three-port deep artificial neural network (ANN), called the *measurement ANN* (*m-ANN*), to capture the link between scene, metasurface coding pattern, and microwave measurement. Being sharply different from the method in the previous section, where only the data processing is treated with the ANN, we here jointly optimize the control coding patterns of the metasurface together with the weights of another

ANN (called the *reconstruction ANN, r-ANN* in short) used to extract the desired information from the raw measurements. To this end, we interpret measurement and reconstruction as, respectively, encoding and decoding the relevant scene information (an image or a gesture classification) in or from a measurement space. Thus, we can employ a variational autoencoder (VAE) framework to jointly learn optimal measurement and processing settings with a standard supervised-learning procedure. This strategy drastically reduces the number of necessary measurements, helping us to remarkably improve many critical metrics, for instance, speed, processing burden, and energy consumption. In this section, we will elaborate on VAE and its sensing implementation.

7.1.2 VAE Principle

We discuss the VAE principle for intelligent integrated sensing. In the context of VAE, the measurement procedure can be viewed as an end-to-end process that, given a scene $\boldsymbol{x}$ (image or class label of probed subject), generates a set of measurements $\boldsymbol{y}$ by sampling from a $\mathcal{C}$ controllable conditional distribution $\boldsymbol{y} \sim q_{\mathcal{C}}(\boldsymbol{y}|\boldsymbol{x}, \boldsymbol{\Theta})$, where $\mathcal{C}$ encapsulates all trainable parameters of hardware setting, namely the user-controlled coding pattern of the reprogrammable coding metasurface. This conditional distribution is known as the likelihood in the framework of Bayesian analysis and can be understood as a stochastic measurement model. Basically, the goal of the data processing pipeline is to produce an estimate $\hat{x}$ of the scene $\boldsymbol{x}$, given the measurements $\boldsymbol{y}$. The estimator $\hat{x}$ serves as an inverse action of the measurement process and can be realized with a deep ANN with network weights $\boldsymbol{\Phi}$. Similar to the measurement process, we denote the estimator with a parametric conditional distribution $\hat{\boldsymbol{x}} \sim p(\hat{\boldsymbol{x}}|\boldsymbol{y}, \boldsymbol{\Phi})$. We propose to simultaneously learn the learnable parameters, namely $\mathcal{C}$ and $\boldsymbol{\Phi}$, of both the measurement process and the reconstruction operator in the context of VAE, such as to optimize the whole sensing performance in a specific task. In light of VAE, the optimal choices of $\mathcal{C}$ and $\boldsymbol{\Phi}$ can be achieved by minimizing the following objective function:

$$\mathscr{L}(\mathcal{C}, \boldsymbol{\Phi}) = -\mathbb{E}_{q_{\mathcal{C}}(\boldsymbol{y}|\boldsymbol{x},\boldsymbol{\Theta})}[\log p(\boldsymbol{x}|\boldsymbol{y}, \boldsymbol{\Phi})] + \mathrm{KL}(q_{\mathcal{C}}(\boldsymbol{y}|\boldsymbol{x}, \boldsymbol{\Theta})||p(\boldsymbol{y})). \tag{7.1}$$

The first term, $-\mathbb{E}_{q_{\mathcal{C}}(\boldsymbol{y}|\boldsymbol{x},\boldsymbol{\Theta})}[\log p_{\boldsymbol{\Phi}}(\boldsymbol{x}|\boldsymbol{y})]$, can be interpreted as the "reconstruction error" of our VAE: it is the log-likelihood of the true latent data given the inferred latent data. The second term, $\mathrm{KL}(q_{\mathcal{C}}(\boldsymbol{y}|\boldsymbol{x}, \boldsymbol{\Theta})|p(\boldsymbol{y}))$, acts as a regularizer and encourages the distribution of the decoder to be close to a chosen prior distribution $p(\boldsymbol{y})$. Both the analog encoder $q_{\mathcal{C}}(\boldsymbol{y}|\boldsymbol{x}, \boldsymbol{\Theta})$ and the digital decoder $p_{\boldsymbol{\Phi}}(\boldsymbol{x}|\boldsymbol{y})$ are treated with two deep ANNs, namely *m-ANN* and *r-ANN*, respectively. Note that the mathematic expectation in the first term of the right-hand side of **Eq. (7.1)** is taken over the distribution $q_{\mathcal{C}}(\boldsymbol{y}|\boldsymbol{x},\boldsymbol{\Theta})$, which embodies the $\mathcal{C}$-controllable measurements and the reconstruction network as a whole. In principle, this term acts as a likelihood, which is used to measure the reconstruction error over $q_{\mathcal{C}}(\boldsymbol{y}|\boldsymbol{x},\boldsymbol{\Theta})$. In contrast, the second term is characterized by KL-divergence, which, as a regularizer, encourages the measurement distribution $q_{\mathcal{C}}(\boldsymbol{y}|\boldsymbol{x},\boldsymbol{\Theta})$ to be close to a chosen prior $p(\boldsymbol{y})$. Here, $p(\boldsymbol{y})$ is chosen to be a zero-mean Gaussian distribution with maximum Shannon information entropy. As such, each measurement is optimized to capture as much information of the probed scene as possible. For the purpose of numerical implementations, $-\mathbb{E}_{q_{\mathcal{C}}(\boldsymbol{y}|\boldsymbol{x},\boldsymbol{\Theta})}[\log p(\boldsymbol{x}|\boldsymbol{y},\boldsymbol{\Phi})]$ is replaced by finite-sample statistical mean approximation over the training dataset. As for the treatment of $\mathrm{KL}(q_{\mathcal{C}}(\boldsymbol{y}|\boldsymbol{x},\boldsymbol{\Theta})||p(\boldsymbol{y}))$, it can be analytically approximated with the Gaussian assumption. In addition, Eq. (7.1) is slightly modified as

$$\mathcal{L}(\mathcal{C},\boldsymbol{\Phi}) = -\mathbb{E}_{q_{\mathcal{C}}(\boldsymbol{y}|\boldsymbol{x},\boldsymbol{\Theta})}[\log p(\boldsymbol{x}|\boldsymbol{y},\boldsymbol{\Phi})] + \gamma \mathrm{KL}(q_{\mathcal{C}}(\boldsymbol{y}|\boldsymbol{x},\boldsymbol{\Theta})||p(\boldsymbol{y})). \tag{7.2}$$

Here, γ is introduced to trade off the contributions from the data misfit and the prior-based regularization.

To determine the optimal settings of $\mathcal{C}$ and $\boldsymbol{\Phi}$, a so-called alternatively iterative approach is applied, as shown in **Fig. 7.2**. Starting with some initializations of $\mathcal{C}$ and $\boldsymbol{\Phi}$, we calculate $\mathcal{C}$ (resp. $\boldsymbol{\Phi}$) for $\boldsymbol{\Phi}$ (resp. $\mathcal{C}$) updated in the last iteration step, followed by calculating $\boldsymbol{\Phi}$ (resp. $\mathcal{C}$) based on the obtained $\mathcal{C}$ (resp. $\boldsymbol{\Phi}$). This procedure is repeated until a stopping criterion is fulfilled. Such an optimization can be implemented using error-backpropagation routines for continuous variables like $\boldsymbol{\Phi}$ and $\boldsymbol{\Theta}$. Apparently, the optimization with respect to $\boldsymbol{\Phi}$ and $\boldsymbol{\Theta}$ can be efficiently realized with the well-known back-propagation (BP) algorithm. However, it is really challenging to minimize **Eq. (7.2)** with respect to the binary control coding sequences $\mathcal{C}$ since it involves a NP-hard combinatorial optimization problem. To address this difficulty, the randomized simultaneous perturbation stochastic approximation (r-SPSA) is slightly modified for this problem. This heuristic optimization approach relies on two randomized descent strategies.

7.1.3 In Situ Imaging of the Human Body

Here, we experimentally evaluate the performance of the preceding VAE-based intelligent integrated sensing system by considering the in situ imaging of the human body in an indoor environment. To that end, we consider the tasks of (i) imaging and (ii) gesture recognition, two representative examples with real-life application in security screening and human–machine interaction. First, we apply our learned EM sensing system to the task of in situ

(a)

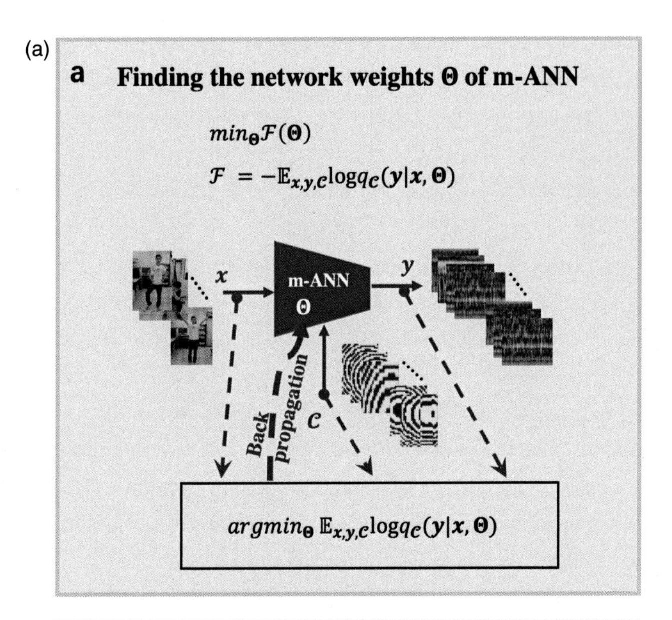

(b)

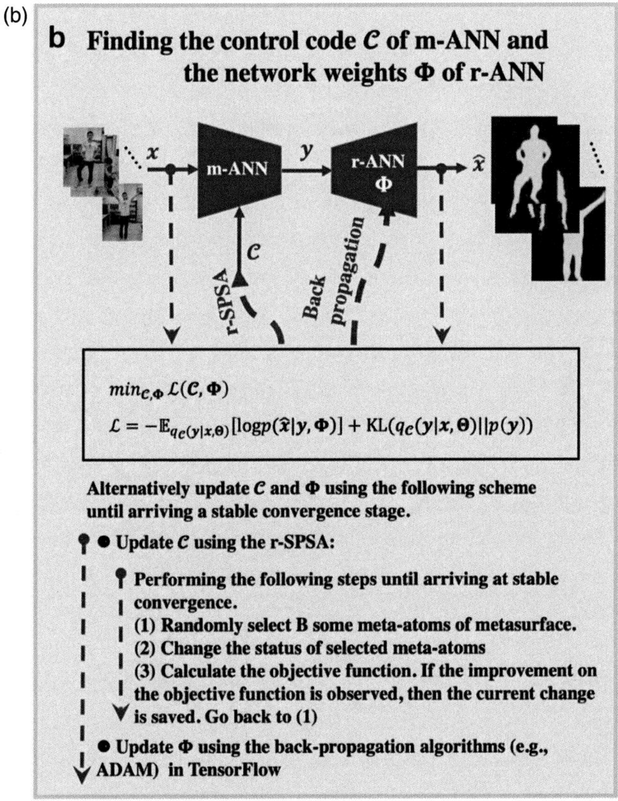

Figure 7.2 (a) The flow chart of learning the network weights of the m-ANN. **(b)** The flow chart of determining the optimal settings of the coding pattern of the metasurface and the network weights of the r-ANN.

high-resolution imaging of a human body in our laboratory environment. For convenience of discussion, several helpful notations are introduced. For all the experiments throughout the section, the *learned measurement* refers to the setting where the control coding pattern of the programmable metasurface is trained alone and the reconstruction network is fixed; the *learned reconstruction* refers to the setting where the reconstruction network is trained alone, and the control coding pattern of the programmable metasurface is fixed; the *learned measurement-reconstruction* refers to the setting where the control coding pattern of the metasurface is jointly trained with the reconstruction network.

As for the learned measurement-reconstruction, we jointly optimize the coding patterns C of the *m-ANN* together with the weights $\mathbf{\Phi}$ of *r-ANN* for the specific task of human body imaging. The integrated ANN, containing a multitude of nonlinear ML layers, is trained with a standard supervised learning procedure. To illustrate the significant improvement of the learned sensing strategy on the image quality over conventional learning-based sensing methods, we consider a two-stage training procedure. During the first stage, the coding patterns of the *m-ANN* and the digital weights of the *r-ANN* are optimized separately. The coding patterns of *m-ANN* are assigned following the two most common state-of-the-art approaches that correspond to using random or PCA-based scene illuminations. During the second stage, *m-ANN* and *r-ANN* were *jointly* trained to achieve the overall optimal sensing performance. In this two-stage training, the benefit reaped by the proposed sensing strategy over the conventional methods can be clearly demonstrated. In our study, we used several people, called training persons in short, to train our intelligent microwave sensing system, and used a different person, called a test person, to test it. In addition, we used 1000 random codes and 1000 PCA-based codes (200 standard PCA-based codes and their 800 perturbations) as raining samples for training $\mathbf{\Theta}$. The ground truth is defined using binarized optical images of the scene. The training of the complex-valued weights of *m-ANN* and *r-ANN* is performed using the Adam optimization method [116]. In addition, the deep residual CNN architecture [117] was used to model the r-ANN. The complex-valued weights were initialized by random weights with a zero-mean Gaussian distribution of standard deviation 10^{-3}. The training was performed on a workstation with an Intel Xeon E5-1620v2 central processing unit, NVIDIA GeForce GTX 1080Ti, and 128 GB access memory. The machine learning platform Tensor Flow was used to design and train the networks in the learned EM sensing system. It took about 267 hours to train the whole learnable sensing pipeline including the m-ANN and r-ANN. Once the r-ANN is well trained, its calculation costs less than 0.6 ms.

Figure 7.3 displays the cross-validation errors over the course of the training iterations for different numbers M of coding patterns of the metasurface (3, 9, 15, and 20). The two stages of the aforementioned training protocol can be clearly distinguished. Since the trainable physical ($\mathcal{C}$) and digital ($\mathbf{\Phi}$) parameters were initialized randomly before training, except for PCA-based $\mathcal{C}$, we have conducted 500 realizations in order to remove any sensitivity to the choices of random initializations made for *m-ANN* and *r-ANN*. In order to evaluate the contribution of the joint training of the measurement setup and data processing pipeline, we have designed a two-stage experiment. First, we trained the reconstruction network alone while fixing the measurement setting (i.e., the programmable metasurface with prespecified control coding patterns). Second, we used a preconvergence checkpoint of the reconstruction network as a starting point for the joint training. At this stage, both the reconstruction network and the measurement setting were jointly trained.

Figure 7.4 reports several selected image reconstruction results of the test person with different body gestures using the aforementioned sensing methods. In line with [1], it can be observed that the sensing quality (here image quality) achieved by jointly optimizing physical ($\mathcal{C}$) and digital ($\mathbf{\Phi}$) parameters is significantly better than if only $\mathbf{\Phi}$ is optimized. This may be intuitively expected since more trainable parameters are available and all a priori knowledge is used in the learned sensing scheme. Experimental results demonstrate that simultaneous learning of measurement and reconstruction settings is remarkably superior to the conventional sensing strategies in which measurement and/or reconstruction are optimized separately (if optimized at all). The benefits of learned sensing are especially strong when the number of measurements is highly limited such that learned sensing enables a remarkable dimensionality reduction. Ultimately, these superior characteristics are enabled by training a unique integrated sensing chain, making use of all available a priori knowledge about the probed scene, task, and constraints on measurement setting and processing pipeline.

To summarize, we discuss intelligent integrated sensing in the context of VAE, which has three critical factors: (i) a reprogrammable coding metasurface is used to shape scene illuminations on the physical layer; (ii) two dedicated deep ANNs are introduced – the *m-ANN* for smart data acquisition and the *r-ANN* for instant data processing; (iii) the learnable weights of the measurement process and those of the processing layer are jointly trained. The presented intelligent sensing strategy with simultaneous learning of measurement and reconstruction procedures yields a superior performance compared to conventional sensing strategies that optimize measurement and processing separately (or not at all). The performance improvements are particularly large when the

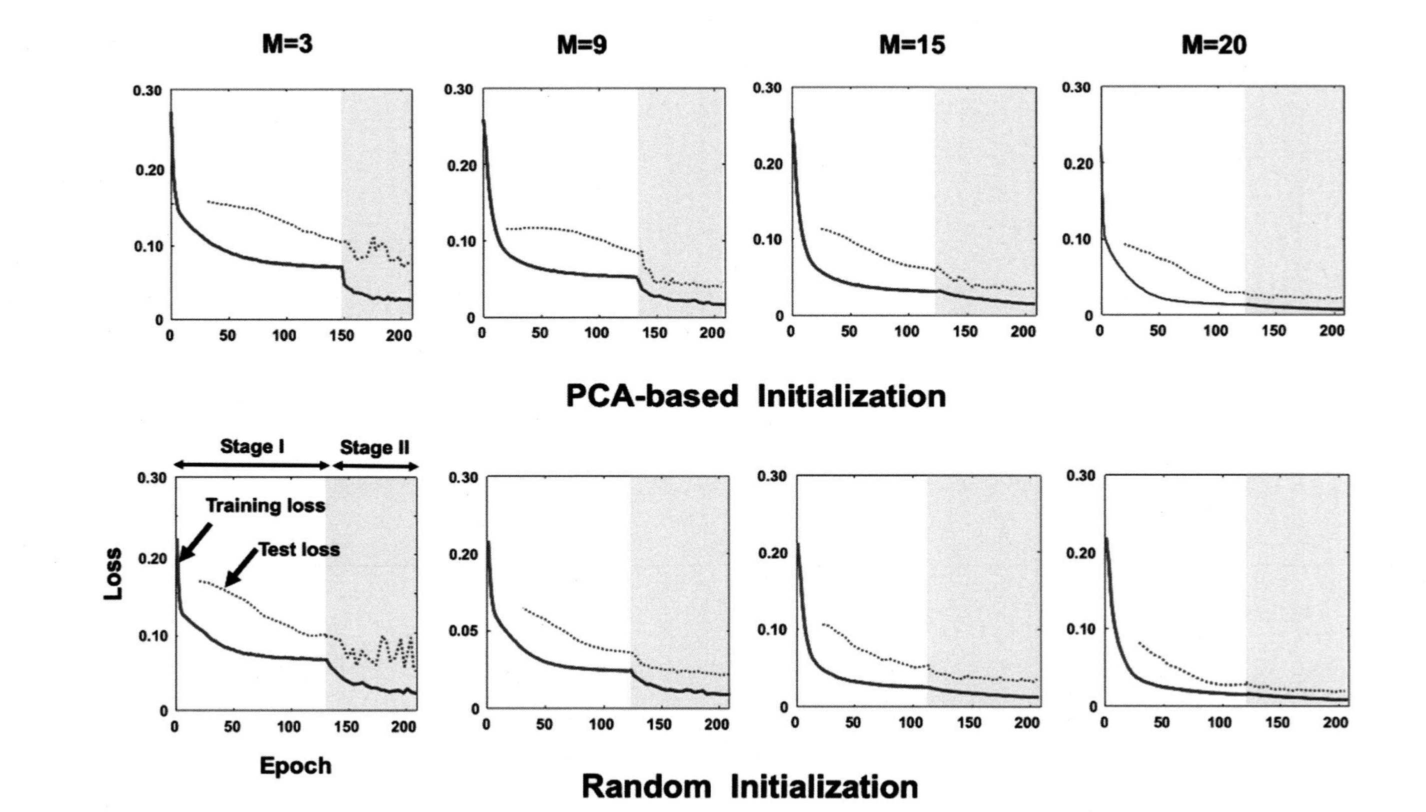

Figure 7.3 Training dynamics for learned EM sensing applied to an imaging task. The dependence of the training and test loss functions on the progress of iterative epochs is shown for different numbers of coding patterns M, namely $M = 3, 9, 15$, and 20. The continuous lines indicate the training loss, and the dashed lines indicate the test loss. The control coding patterns of the metasurface are initialized randomly (**a**) or PCA-based (**b**).

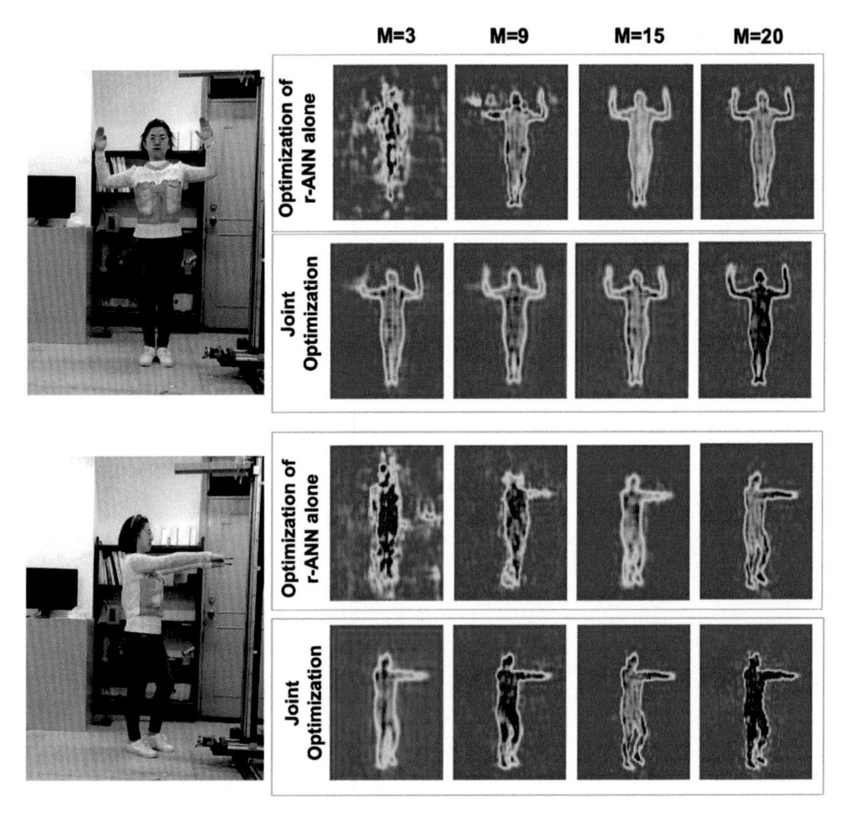

Figure 7.4 Experimental results for learned EM sensing applied to an imaging task. For three different poses, we display the images reconstructed with different numbers of coding patterns of the metasurface, M, for the case of only optimizing $\mathbf{\Phi}$ (first row) or jointly optimizing C and $\mathbf{\Phi}$ (second row). In this set of experiments, the random initialization is used.

number of measurements is very limited, as we have experimentally demonstrated. Our work paves the path to low-latency microwave sensing for security screening, human–computer interaction, health care, and automotive radar.

7.2 Free-Energy-Based Intelligent Integrated Sensing

In this section, we discuss the free-energy-based intelligent integrated sensing system proposed by Wang et al. [118–121] Similar to the system discussed in the previous section, such a sensing system relies on two critical factors: the joint optimization of data acquisition and data processing, and the synergetic exploitation of artificial materials (i.e., reprogrammable metasurfaces) for adaptive data acquisition on the physical level and artificial intelligence (deep learning) for instant data processing on the digital level. Here, we present real-world experimental results selected from [109] to demonstrate that such an intelligent system enables us to see clearly human behaviors behind a 60-cm-thick reinforced concrete wall with the frame rate on the order of tens of Hz. We here mean by the real-world setting that the target is in a really complicated indoor physical environment, and acquired signals are seriously disturbed by unknown cochannel interferences. Here, we remark that the frame rate can be achieved on the order of tens of kHz in principle if more specialized transceivers are used rather than commercial radio devices (i.e., USRP), as pointed out in [109].

7.2.1 System Design

The intelligent integrated sensing system, working at 2.4 GHz, is a software-defined system in favor of the high-frame-rate EM sensing, which was used for monitoring human behaviors in an indoor environment. As shown in **Fig. 7.5a**, such a system consists of a one-bit reprogrammable coding metasurface, a commercial software-defined radio device (Ettus USRP X310), a transmitting antenna, a three-antenna receiver, and a personal computer. Here, the one-bit reprogrammable coding metasurface is designed to be composed of 3×3 identical metasurface panels, and each panel has 8×8 meta-atoms with a size of 54×54 mm^2 (see **Fig. 7.5c**). The one-bit reprogrammable coding metasurface controlled with ANNs is utilized for two purposes: (i) manipulating adaptively the EM wave fields toward the target, suppressing the unwanted disturbances from surrounding environment like walls, furnishings, and so on; (ii) serving as an electronically controllable coding aperture in a compressive-sensing manner. Owing to the large-view field enabled by the large-aperture reprogrammable metasurface, the target's information can be readily captured by two fixed receivers. Therefore, it can be expected that the target's information can be readily retrieved from compressive measurements. Both the USRP and the metasurface are communicated with the host computer via Ethernet under the transmission control protocol (TCP); meanwhile, the USRP has I/O series communication with the one-bit reprogrammable coding metasurface. The host computer is responsible for calculating the control patterns and sends these patterns to the metasurface through the FPGA module; at the same time, it sends a command signal to the USRP for synchronizing its transmitting and receiving channels, as shown in **Fig. 7.5b**. To trade off the imaging quality with efficiency, we explore 18 patterns for compressive microwave measurement per image in this work. For each control pattern, the USRP under the control of the host computer will generate the radio signal with a chirp waveform, radiate it into the investigation domain through the transmitting antenna, and receive the

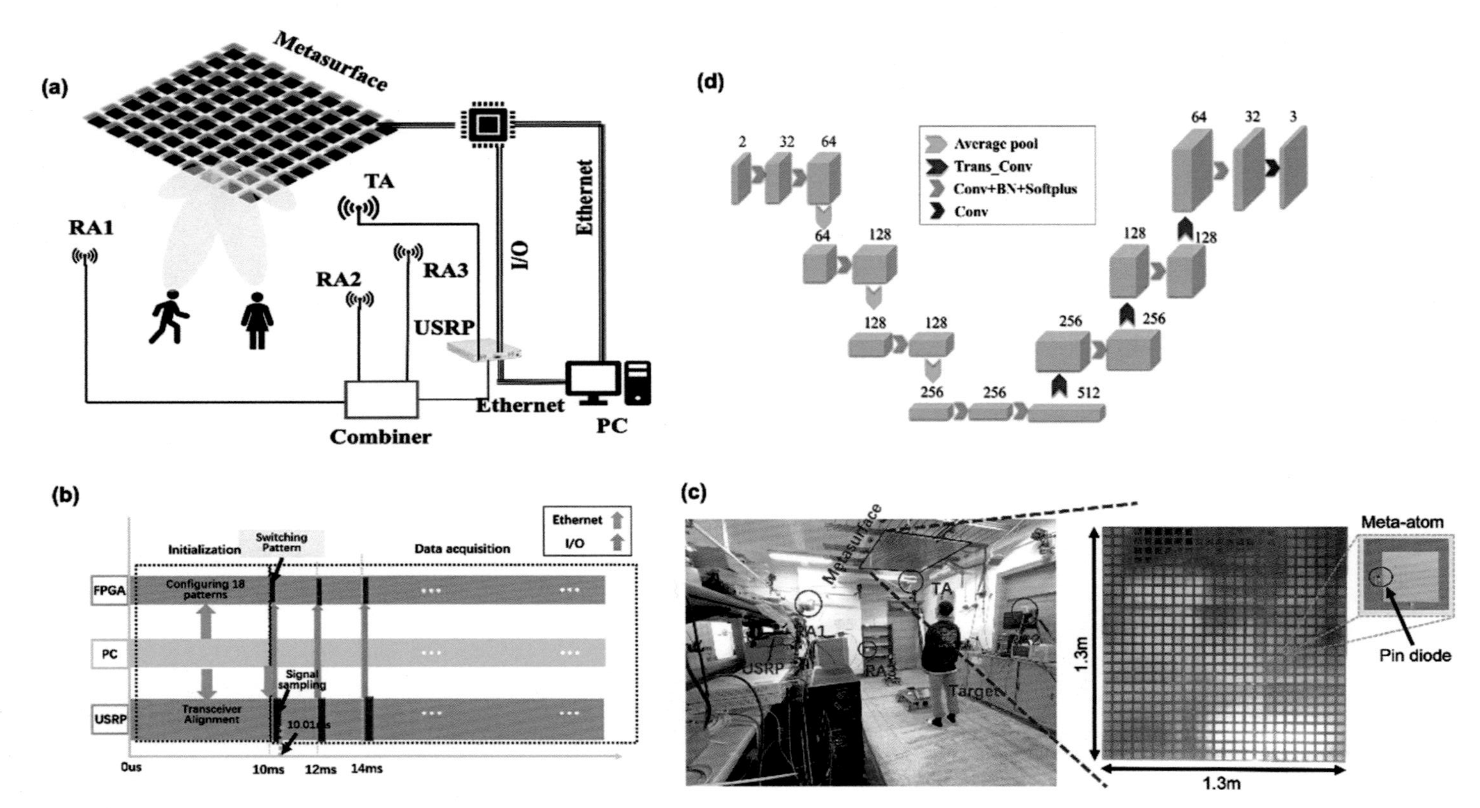

Figure 7.5 Configuration of the metasurface-based intelligent integrated sensing system working at around 2.4 GHz. **(a)** The sketch map of the proposed intelligent EM metasurface camera system. **(b)** The operational procedure of data acquisition of the proposed EM camera. **(c)** The experimental setup in an indoor environment. **(d)** The proposed U-net network along with necessary parameters.

echoes reflected from the target. It will take about 36 ms to produce one microwave image, implying the frame rate achievable is about 27 Hz. Afterwards, the acquired echoes are processed by artificial neural networks in the host computer, which is directly responsible for the object reconstruction and recognition.

7.2.2 Principle of Free-Energy Minimization

We discuss the free-energy principle for the metasurface-based intelligent integrated sensing system. We assume that, for the target's state s_t at time t, the metasurface camera aims at organizing the measurement strategy π_t, collecting measurements o_t, and retrieving the target. To build a machine model for this problem, we explored the probabilistic generative model and its Bayesian inference solution. For the target-sensor scenario with the generative distribution $P_\varphi(s_{\le t}, o_{\le t}|\pi_{\le t})$, the intelligent metasurface camera has a picture of it, which is characterized with a posterior distribution (i.e., recognition function) $Q_\theta(s_{\le t}|o_{\le t}, \pi_{\le t})$. Here, φ and θ encapsulate all trainable parameters defining P_φ and Q_θ, respectively. Then, the generative network P_φ and the inference network Q_θ could be achieved by minimizing the following free energy [15]:

$$\mathcal{F} = \mathbb{E}_{Q_\theta(s_{\le t}|o_{\le t},\pi_{\le t})}\left[\ln\left(\frac{Q_\theta(s_{\le t}|o_{\le t}, \pi_{\le t})}{P_\varphi(s_{\le t}, o_{\le t}|\pi_{\le t})}\right)\right]. \tag{7.3}$$

Under well-known Markov chain approximation, Eq. (7.3) can be expressed as

$$\mathcal{F} = \sum\nolimits_{t=1}^{T} J_t \tag{7.4}$$

where

$$J_t = J_{t-1} - \mathbb{E}_{Q_\theta(s_{\le t-1}|o_{\le t-1},\pi_{\le t-1})}\left[\underbrace{\mathbb{E}_{Q_\theta(s_t|o_t,s_{\le t-1},\pi_t)}\ln\Big(P_\varphi(o_t|s_t, \pi_t)\Big)}_{\text{Likelihood}}\right]$$

$$-\mathbb{E}_{Q_\theta(s_{\le t-1}|o_{\le t-1},\pi_{\le t-1})}\left[\underbrace{\mathbb{E}_{Q_\theta(s_t|o_t,s_{\le t-1},\pi_t)}\ln\left(\frac{P_\varphi(s_t|s_{\le t-1})}{Q_\theta(s_t|o_t, s_{\le t-1}, \pi_t)}\right)}_{\text{KL divergence}}\right].$$

Note that $\mathbb{E}_{Q_\theta(s_t|o_t,s_{\le t-1},\pi_t)}\ln\Big(P_\varphi(o_t|s_t, \pi_t)\Big)$ describes the likelihood or observation accuracy at time t, while $\mathbb{E}_{Q_\theta(s_t|o_t,s_{\le t-1},\pi_t)}\ln\left(\frac{P_\varphi(s_t|s_{\le t-1})}{Q_\theta(s_t|o_t,s_{\le t-1},\pi_t)}\right)$, Kullback–Leibler (KL) distance, reflects the recognition complexity of $Q_\theta(s_t|o_t, s_{\le t-1}, \pi_t)$. In addition, we assume $Q_\theta(s_t|o_t, s_{\le t-1}, \pi_t) = \mathcal{N}(s_t|f_\theta(o_t, \pi_t), \alpha^2 I)\mathcal{N}(s_t|s_{t-1}, \beta^2 I)$, where the nonlinear function f_θ is modeled with a U-net artificial neural network (see **Fig. 7.5d**), and α^2 and β^2 are two trainable parameters. Now, one can determine the measurement strategy π_t, the generative network P_φ, and the inference network Q_θ by minimizing **Eq. (7.4)**.

Before closing this section, more details about the U-net network [122] modeling f_θ are discussed here. The U-net network utilized here has double-channel input: one channel is from the real part of the preprocessed microwave signal, the other is from the imaginary part. Features of microwave signals are extracted layer by layer and then gradually approach the labeled IUV three-channel images [123]. The training is performed over a GPU computer with a single Nvidia GTX2080Ti, and the training setup is made: the optimizer is Adam [116], the learning rate is 10^{-3}, the weight decay rate is 5×10^{-5}, and the batch size is 128. Although the whole investigation domain is really big, the target occupies only a small fraction of it. Taking this observation into account, we propose to enforce the values of the pixels outside the target to be zero each iteration during the training procedure, which is referred to as the window-Adam for convenience in notation.

7.2.3 Through-the-Wall Sensing Results

We demonstrate the performance of the preceding system for the in situ imaging of a human freely acting in our lab. In order to train this system, we integrated a commercial optical binocular camera ZED2 from Stereolabs and synchronized it with the EM sensing system through the host computer. Then, the optical videos by the ZED2 are utilized as the training samples after a sequence of processes including background removal, segmentation, and IUV-transformation. We collected 8×10^4 pairs of labeled training videos and took around 18 hours to train the metasurface-based intelligent integrated sensing system. The metasurface-based intelligent sensing system, once trained, can produce a high-fidelity video with a frame rate of about 20 Hz.

Signal Model and Data Preprocessing The chip signal waveform transmitted by the USRP reads

$$s(t) = \exp\left(j\left(2\pi f_c t + \pi K t^2\right)\right), \quad 0 \le t \le T, \tag{7.5}$$

where $j = \sqrt{-1}$, $f_c = 2.424$ GHz is the carrier frequency, $K = B/T$ denotes the sweep rate of the chirp, $B = 50$ MHz is the frequency bandwidth, and $T = 10$ μs is the chirp pulse duration. Assuming that a transmitter at $\boldsymbol{r}_T$ gives rise to a frequency-domain signal $s(\omega)$, and a point-like object with reflection coefficient $\sigma(\boldsymbol{r}_o)$ is situated at $\boldsymbol{r}_o$, where ω denotes the angular frequency. Such a pointlike target model makes sense, and the following discussion can be readily extended for the case of extended objects in terms of the linear supposition principle in the context of Born scattering approximation [24]. Then, the echo acquired by the receiver at $\boldsymbol{r}_R$ can be approximated as

$$\begin{aligned} y(\boldsymbol{r}_R;\omega,\mathcal{C}) &\approx s(\omega)\sigma(\boldsymbol{r}_o)g(\boldsymbol{r}_R,\boldsymbol{r}_o;\omega)g(\boldsymbol{r}_T,\boldsymbol{r}_o;\omega) \\ &+ s(\omega)\sigma(\boldsymbol{r}_o)g(\boldsymbol{r}_R,\boldsymbol{r}_o;\omega)\left(\sum\nolimits_n \Gamma_n^{\mathcal{C}}(\omega)g(\boldsymbol{r}_T,\boldsymbol{r}_n;\omega)g(\boldsymbol{r}_o,\boldsymbol{r}_n;\omega)\right) \\ &+ s(\omega)g(\boldsymbol{r}_R,\boldsymbol{r}_T;\omega) + s(\omega)\sum_n \Gamma_n^{\mathcal{C}}(\omega)g(\boldsymbol{r}_R,\boldsymbol{r}_n;\omega)g(\boldsymbol{r}_T,\boldsymbol{r}_n;\omega) + \varepsilon(\omega). \end{aligned} \tag{7.6}$$

Herein, $g(\boldsymbol{r}_R,\boldsymbol{r}_o;\omega)$ denotes the so-called Green's function of considered physical environment, which characterizes the system response at $\boldsymbol{r}_R$ given a radio source at $\boldsymbol{r}_o$. $\Gamma_n^{\mathcal{C}}(\omega)$ represents the reflection coefficient of the nth meta-atom at $\boldsymbol{r}_n$, when the metasurface is configured with the control coding pattern $\mathcal{C}$. Note that the first and second terms in the right-hand side of **Eq. (7.6)** carry the target's information, while the other terms are usually target-independent. It is trivial to remove the third and fourth terms by exploring a simple background removal operation. It is noted that $\varepsilon(\omega)$ accounts for disturbances from the aforementioned in-band inferences, environment clutters, and system noise. As mentioned previously, the intelligent system working at around 2.4 GHz is deployed in a real-world indoor environment. In this scenario, the acquired signals are inevitably disturbed by unwanted but unknown in-band wireless signals everywhere, and there are plenty of unwanted interferences arising from surrounding environment such as walls, furniture, and so on. It is challenging due to its statistically nonstationary for real-time demands, since conventional filter-based methods, such as the time-frequency filtering and principal component filtering, are typically computationally prohibitive. In order to resolve this problem, the deep learning strategy was explored. We designed an end-to-end deep convolutional network, termed as Filter-CNN, mapping the noisy signal after background removal to the desired denoised signal. Its training was performed in a GPU-equipped personal computer with Nvidia GTX2080Ti, and major parameters were set as follows: the optimizer is Adam, batch size = 128, initial learning rate = 0.01, and iteration epochs = 100. Such training costs 2 hours. Once the Filter-CNN was trained well, the filtering time for a group of 20×1000 data was about 0.3 s, while conventional methods needed at least 11 s.

Figure 7.6b and e presents the results for the denoised signals, where 18 random control coding patterns of programmable metasurface are used, and a human target stands quietly in the outdoor environment. For comparison, corresponding down-converted signals are also provided in **Fig. 7.6a and d**. It

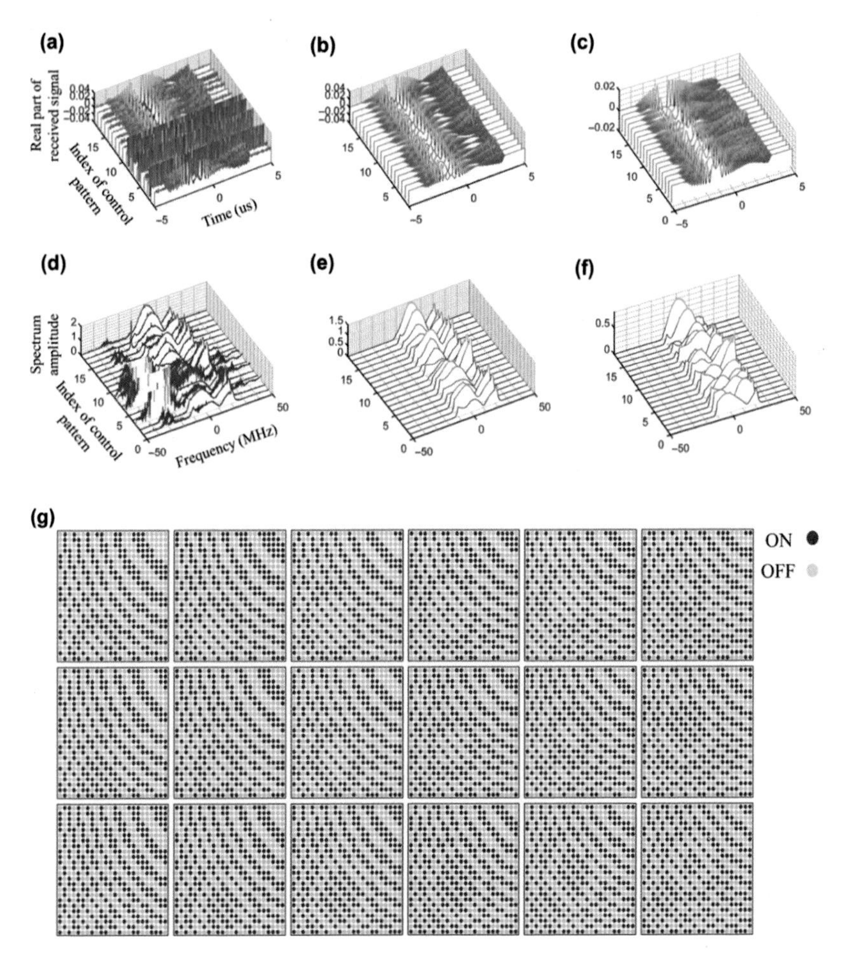

Figure 7.6 Experimental results of the signal denoise for the proposed intelligent EM metasurface camera working at around 2.4 GHz. **(a)** Real parts of time-domain down-converted signals where 18 random control patterns of metasurface are considered. This figure clearly shows very serious in-band and out-of-band disturbances. **(b)** Real parts of 18 denoised time-domain signals in **a** through the proposed Filter-CNN. **(c)** Real parts of mean-valued-filtered signals of **b**. **(d–f)** are the spectrum amplitudes corresponding to those in **a–c**. **(g)** Eighteen random control patterns of the metasurface used in this set of experiments.

can be clearly observed from **Fig. 7.6a–b and d–e** that the overwhelming number of unwanted inferences can be well filtered out using our Filter-CNN. Recall **Eq. (7.6)**; the first term characterizes the direct arrival from the source to the receiver, and thus is out of the control of the programmable metasurface. To demonstrate the effect of the metasurface on the compressive measurements, the first term in **Eq. (7.6)** is removed by a mean-value filter with respect to the slow time (i.e., measurement index), and corresponding results are plotted in **Fig. 7.6c and f**. From these figures, we can see that the reprogrammable coding metasurface can flexibly manipulate the acquisition of the microwave signals carrying the target's information, implying that the target's information can be efficiently captured by a fixed receiver. Additionally, 18 random control coding patterns of the metasurface involved in **Fig. 7.6a–f** have been plotted in **Fig. 7.6g**. Recall **Eq. (7.6)**; one interesting conclusion can be observed, namely the good measurements imply the acquired signals have a good signal-to-noise ratio (SNR), which can be achieved by controlling the coding patterns of the one-bit reprogrammable coding metasurface. Intuitively, such a measurement strategy can be conceived by designing the control pattern of the metasurface such that the resultant radiation beams are focused toward the target, where the prior on the target's location can be estimated from the image obtained at the last time.

In Situ Imaging Results We show the through-wall sensing performance of the developed sensing system. In this setting, the target freely acts in a corridor outside our lab with a 60-cm-thick load-bearing concrete wall. **Figure 7.7** reports through-wall IUV images at selected moments. It can be observed from this set of figures that the image quality in an outdoor environment is comparable to those in an indoor environment, and that the actions of the test person behind a 60-cm-thick concrete wall remain to be clearly identified. To further demonstrate the benefit enabled by the intelligence, we conducted a set of experiments with random control coding patterns of the metasurface, and no acceptable results have been obtained due to the weak measurements. Therefore, we can conclude that the proposed EM metasurface camera with the intelligence has better learning ability than that without the intelligence.

Finally, we further consider the effect on sensing performance from the use of the window Adam optimizer, where the adaptive focusing measurement is considered. **Figure 7.8a and b** presents the IUV images of a test person with four different gestures with growth of epochs when the Adam and window Adam optimizers are used, respectively. **Figure 7.8c** compares the training convergences when the window operation in the Adam optimizer is used or not. **Figure 7.8d** compares quantitatively the PSNR's histogram of test results, showing the benefits of the window-Adam optimizer relative to the conventional Adam optimizer. From the preceding results, several conclusions can be drawn. First, the window operation in the Adam optimizer is really helpful in improving the training efficiency. Notably, during the training process, the location knowledge of the target can be rapidly learned by our EM camera,

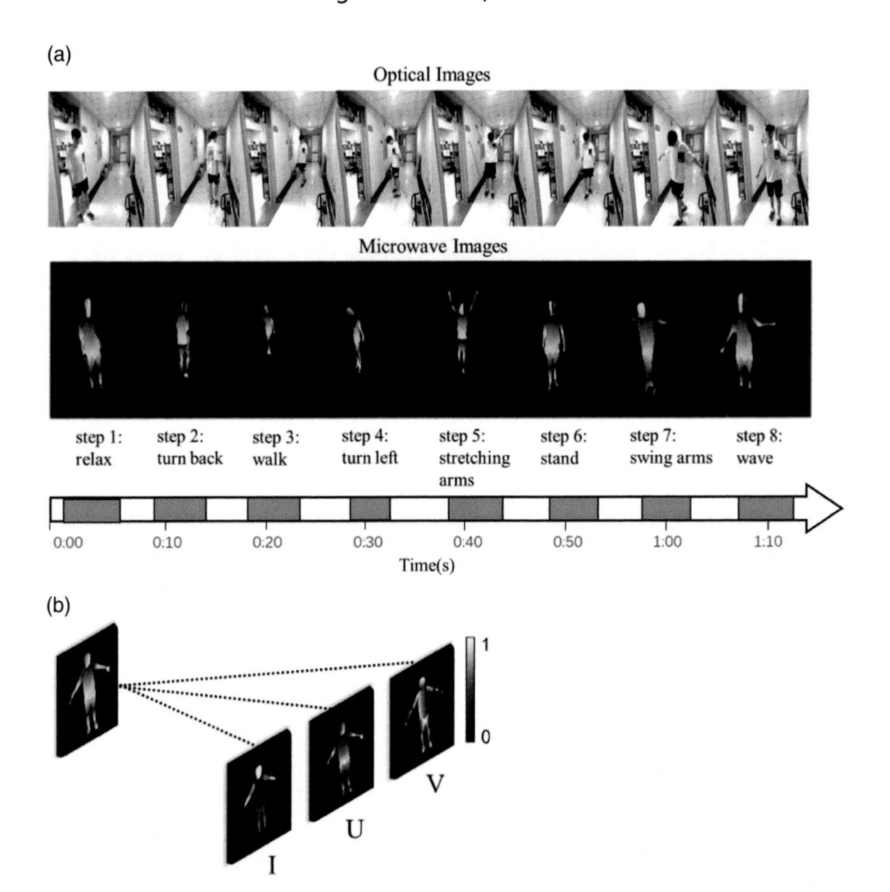

Figure 7.7 (a) Experimental in situ imaging results of the test person freely acting in a corridor outside our lab with a 60-cm-thick load-bearing concrete wall. (Top) RGB images recorded by an optical ZED2 camera at selected moments. (Middle) IUV images by our intelligent EM metasurface camera at selected moments corresponding to those in the top row. (Bottom) Time line. **(b)** An IUV image has three channels: I-channel, U-channel, and V-channel. The I-channel image is the classification of pixels that belong to either background or different parts of body, which provide a coarse estimate of surface coordinates. The UV channels provide the result of mapping all human pixels of a RGB image to the 3D surface of the human body.

meanwhile reducing ghosting, blurring, and other distortions. Second, the proposed intelligent EM camera enables us to see clearly human behaviors in a complicated visually invisible environment.

To summarize, we discuss the metasurface-based intelligent integrated sensing system and its operational free-energy principle. The sensing system relies on two critical parts: (1) a large-aperture programmable metasurface for

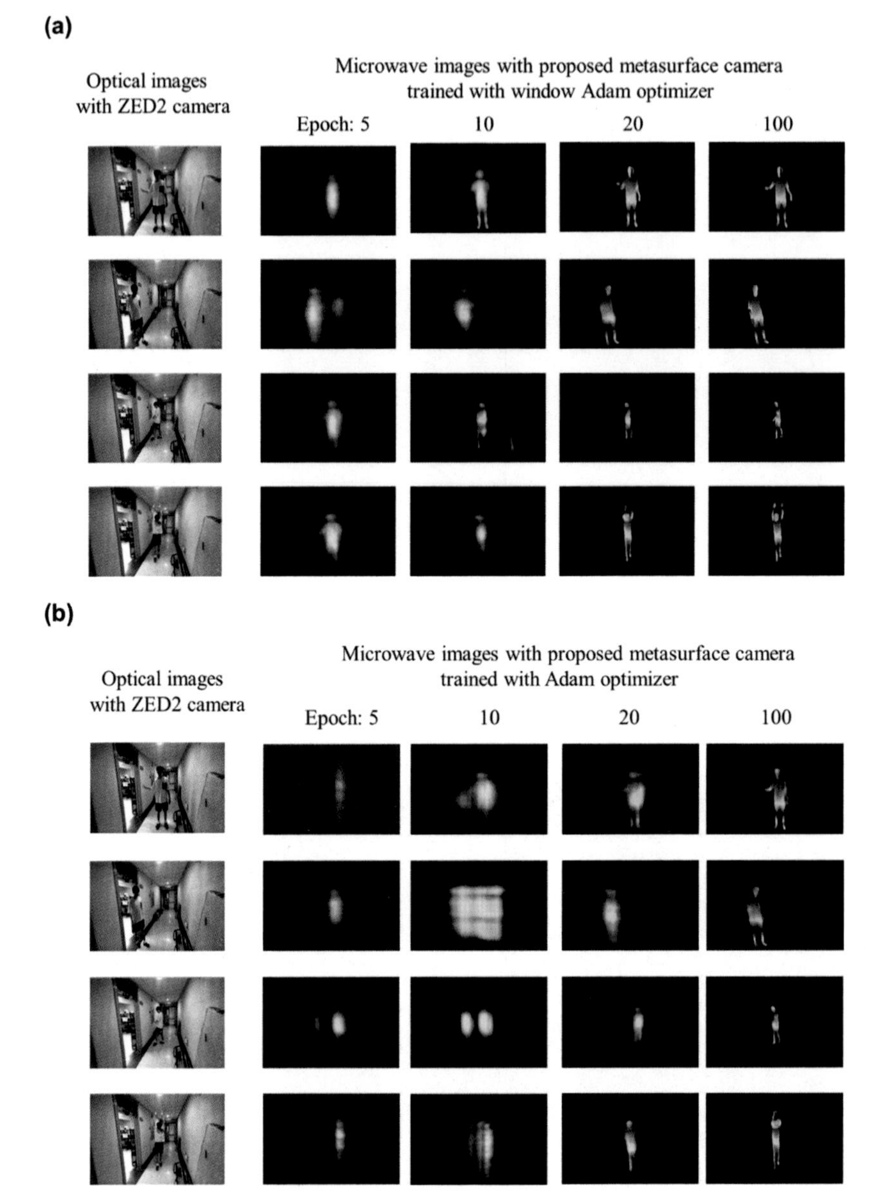

Figure 7.8 Performance evaluation for sensing human behaviors in through-wall environment. **(a)** and **(b)** correspond to selected IUV images of the test person with four different gestures with varying training epochs, where the window-Adam and Adam are used, respectively. **(c)** Comparison of convergence behaviors of MSEs over 1000 1.2-s-length microwave videos as a function of training epochs when the window-Adam and Adam optimizers are used, respectively. **(d)** Figures are for the normalized histograms of PSNRS over 1000 1.2-s-length microwave videos recorded by using our EM camera trained with the Adam and window-Adam optimizer, respectively.

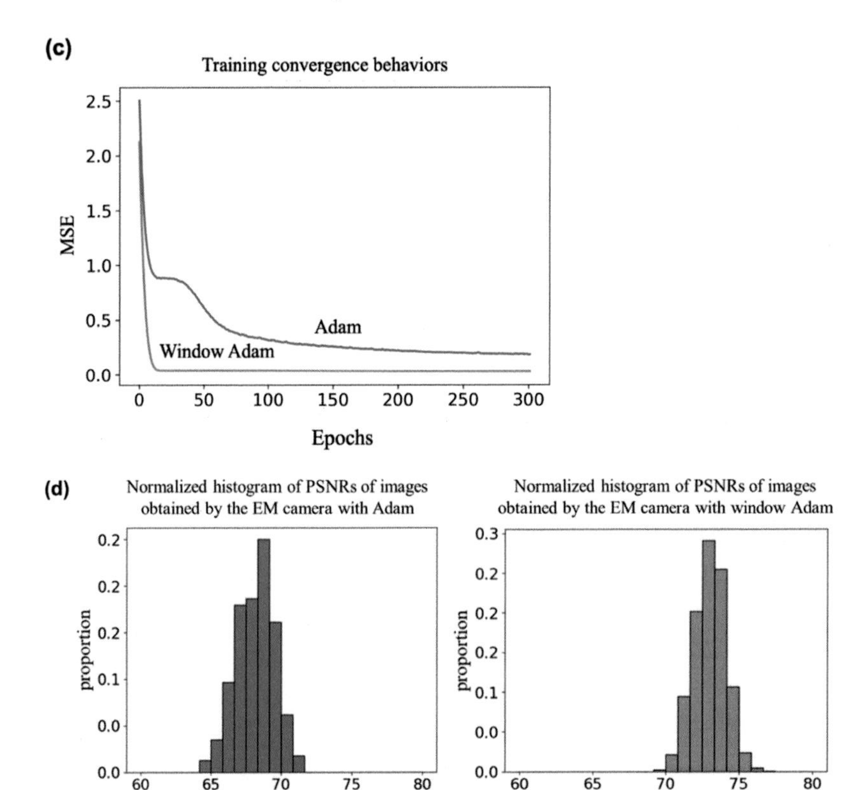

Figure 7.8 (cont.)

adaptive data acquisition; and (2) artificial neural networks for instant data processing. From our discussion, we can see that such an intelligent system enables one to monitor clearly human behaviors behind a 60-cm-thick concrete wall in real-world settings. The presented strategy could open up a promising route toward smart community, Internet-of-Things (IoT) connectivity, and beyond, and can be transposed to other frequencies and other types of wave phenomena.

References

1. T. J. Cui, M. Q. Qi, X. Wan, J. Zhao, and Q. Cheng, Coding metamaterials, digital metamaterials and programmable metamaterials. *Light Sci. Appl.*, **3** (2014), e218.
2. T. J. Cui, Microwave metamaterials: from passive to digital and programmable controls of electromagnetic waves. *J. Optics*, **19** (2017), 084004.
3. L. Li and T. J. Cui, Information metamaterials: from effective media to real-time information processing systems. *Nanophotonics*, **8** (2019), 703–724.
4. L. Zhang, X. Q. Chen, S. Liu et al., Space-time-coding digital metasurfaces. *Nat. Commun.*, **9** (2018), 4334.
5. B. Gholipour, J. Zhang, K. F. MacDonald, D. W. Hewak, and N. I. Zheludev, An all-optical, non-volatile, bidirectional, phase-change meta-switch. *Adv. Mater.*, **25** (2013), 3050–3054.
6. Q. Wang, E. T. F. Rogers, B. Gholipour et al., Optically reconfigurable metasurfaces and photonic devices based on phase change materials. *Nat. Photonics*, **10** (2015), 60–65.
7. G. Kaplan, K. Aydin, and J. Scheuer, Dynamically controlled plasmonic nano-antenna phased array utilizing vanadium dioxide. *Opt. Mater. Express*, **5** (2015), 2513.
8. M. J. Dicken, K. Aydin, I. M. Pryce et al., Frequency tunable near-infrared metamaterials based on VO2 phase transition. *Opt. Express*, **17** (2009), 18330.
9. L. Ju, B. Geng, J. Horng et al., Graphene plasmonics for tunable terahertz metamaterials. *Nat. Nanotechnol.*, **6** (2011), 630–634.
10. J.-Y. Ou, E. Plum, J. Zhang, and N. I. Zheludev, An electromechanically reconfigurable plasmonic metamaterial operating in the near-infrared. *Nat. Nanotechnol.*, **8** (2013), 252–255.
11. Website of Teratech Components Ltd, www.teratechcomponents.com/.
12. W. C. B. Peatman, P. A. D. Wood, D. Porterfield, T. W. Crowe, and M. J. Rooks, Quarter-micrometer GaAs Schottky barrier diode with high video responsivity at 118μm. *Appl. Phys. Lett.*, **63** (1992), 284–290.
13. A. Ghanekar, J. Ji, and Y. Zheng, High-rectification near-field thermal diode using phase change periodic nanostructure. *Appl. Phys. Lett.*, **109** (2016), 123106.
14. A. S. Barker, Jr., H. W. Verleur, and H. J. Guggenheim, Infrared optical properties of vanadium dioxide above and below the transition temperature. *Phys. Rev. Lett.*, **17** (1966), 1286–1289.

15. Y. B. Li, L. L. Li, B. B. Xu et al., Transmission-type 2-bit programmable metasurface for single-sensor and single-frequency microwave imaging. *Sci. Rep.*, **6** (2016), 23731.
16. L. L. Li, H. Ruan, C. Liu et al., Machine-learning reprogrammable metasurface imager. *Nat. Commun.*, **10** (2019), 1082.
17. H. Yang, X. Cao, F. Yang et al., A programmable metasurface with dynamic polarization, scattering and focusing control. *Sci. Rep.*, **6** (2016), 35692.
18. Y. Shuang, H. Zhao, W. Ji, T. J. Cui, and L. Li, Programmable high-order OAM-carrying beams for direct-modulation wireless communications. *IEEE J. Emerg.*, **10** (2020), 29–37.
19. J. Q. Han, L. Li, H. Yi, and Y. Shi, 1-bit digital orbital angular momentum vortex beam generator based on a coding reflective metasurface. *Opt. Mater. Express*, **8** (2018), 3470–3478.
20. D. Zhang, X. Y. Cao, and H. H. Yang, Radiation performance synthesis for OAM vortex wave generated by reflective metasurface. *IEEE Access*, **6** (2018), 28691–28701.
21. X. G. Zhang, Y. L. Sun, Q. Yu et al., Smart Doppler cloak operating in broad band and full polarizations. *Adv. Mater.*, **33** (2021), 2007966.
22. L. Li, T.-J. Cui, W. Ji et al., Electromagnetic reprogrammable coding-metasurface holograms. *Nat. Commun.*, **8** (2017), 197.
23. L. Li, Y. Shuang, Q. Ma et al., Intelligent metasurface imager and recognizer. *Light Sci. Appl.*, **8** (2019), 97.
24. H.-Y. Li, H.-T. Zhao, M.-L. Wei et al., Intelligent electromagnetic sensing with learnable data acquisition and processing. *Patterns*, **1** (2020), 100006.
25. J. Wang, Y. Li, Z. H. Jiang et al., Metantenna: when metasurface meets antenna again. *IEEE Trans. Antennas Propag.*, **68** (2020), 1332–1347.
26. Y. Shuang, L. Li, Z. Wang et al., Controllable manipulation of Wi-Fi signals using tunable metasurface.*J. Radars*, **10** (2021), 313–325.
27. R. W. Gerchberg and W. O. Saxton, A practical algorithm for the determination of the phase from image and diffraction plane pictures. *Optik*, **35** (1972), 227–246.
28. D. Gabor, A new microscopic principle. *Nature*, **161** (1948), 777–778.
29. T.-C. Poon and J.-P. Liu, *Introduction to Modern Digital Holography: with Matlab*. (Cambridge: Cambridge University Press, 2014).
30. G. Zheng, H. Mühlenbernd, M. Kenney et al., Metasurface holograms reaching 80% efficiency. *Nat. Nanotechnol.*, **10** (2015), 308–312.
31. P. Genevet, J. Lin, M. A. Kats, and F. Capasso, Holographic detection of the orbital angular momentum of light with plasmonic photodiodes. *Nat. Commun.*, **3** (2012), 1278.

32. N. Yu, P. Genevet, M. A. Kats et al., Light propagation with phase discontinuities: generalized laws of reflection and refraction. *Science*, **334** (2011), 333–337.
33. S. Larouche, Y.-J. Tsai, T. Tyler, N. M. Jokerst, and D. R. Smith, Infrared metamaterial phase holograms. *Nat. Mater.*, **11** (2012), 450–454.
34. L. Huang, H. Mühlenbernd, X. Li et al., Broadband hybrid holographic multiplexing with geometric metasurfaces. *Adv. Mater.*, **27** (2015), 6444–6449.
35. J. Durnin, J. J. Miceli, and J. H. Eberly, Diffraction-free beams. *Phy. Rev. Lett.*, **58** (1987), 1499–1501.
36. Z. Bouchal, J. Wagner, and M. Chlup, Self-reconstruction of a distorted nondiffracting beam. *Opt. Commun.*, **151** (1998), 207–211.
37. J. Chen, J. Ng, Z. Lin, and C. T. Chan, Optical pulling force. *Nat. Photon.*, **5** (2011), 531–534.
38. A. Dogariu, S. Sukhov, and J. Jose Sáenz, Optically induced "negative forces." *Nat. Photon.*, **7** (2013), 24–27.
39. B. Thidé, H. Then, J. Sjöholm et al., Utilization of photon orbital angular momentum in the low-frequency radio domain. *Phys. Rev. Lett.*, **8** (2007), 087701.
40. F. Tamburini, E. Mari, A. Sponselli et al., Encoding many channels in the same frequency through radio vorticity: first experimental test. *New J. Phys.*, **14** (2013), 3.
41. Y. Ren, L. Li, G. Xie et al., Line-of-sight millimeter-wave communications using orbital angular momentum multiplexing combined with conventional spatial multiplexing. *IEEE Trans. Wirel.*, **5** (2017), 3151–3161.
42. Y. Yan, G. Xie, M. Lavery et al., High-capacity millimetre-wave communications with orbital angular momentum multiplexing. *Nat. Comms.*, **5** (2014), 4876.
43. L. Li and Li Fang, Beating the Rayleigh limit: orbital-angular-momentum-based super-resolution diffraction tomography. *Phys. Rev. E.*, **88** (2013), 033205.
44. S. Liu, T. J. Cui, L. Zhang et al., Convolution operations on coding metasurface to reach flexible and continuous controls of terahertz beams. *Adv. Sci.*, **10** (2016), 1600156.
45. Y. Shuang, H. Zhao, M. Wei et al., One-bit quantization is good for programmable coding metasurfaces. *Sci. China Inf. Sci.*, **65** (2022), 172301.
46. A. Derode, A. Tourin, and M. Fink, Ultrasonic pulse compression with one-bit time reversal through multiple scattering. *J. Appl. Phys.*, **85** (2014), 6434.

47. D. Li, Ergodic capacity of intelligent reflecting surface-assisted communication systems with phase errors. *IEEE Commun. Lett.*, **24** (2020), 1646.
48. P. Xu, G. Chen, Z. Yang, and M. Di Renzo, Reconfigurable intelligent surfaces assisted communications with discrete phase shifts: how many quantization levels are required to achieve full diversity? *IEEE Wirel. Commun. Lett.*, **1** (2020), 358–362.
49. H. Zhao, Y. Shuang, M. Wei, and T.-J. Cui, Metasurface-assisted massive backscatter wireless communication with commodity Wi-Fi signals. *Nat. Commun.*, **11** (2020), 3926.
50. T.-J. Cui, S. Liu, and L.-L. Li, Information entropy of coding metasurface. *Light Sci. Appl.*, **5** (2016), e16172.
51. H. Wu, G. D. Bai, S. Liu et al., Information theory of metasurfaces. *Natl. Sci. Rev.*, **7** (2020), 561.
52. C. E. Shannon, A mathematical theory of communication. *Bell Syst. Tech. J.*, **27** (1948), 379.
53. M. F. Duarte, M. A. Davenport, D. Takhar et al., Single-pixel imaging via compressive sampling. *IEEE Signal Process. Mag.*, **25** (2008), 83–91.
54. C. M. Watts, D. Shrekenhamer, J. Montoya et al., Terahertz compressive imaging with metamaterial spatial light modulators. *Nat. Photonics*, **8** (2014), 605–609.
55. W. L. Chan, K. Charan, D. Takhar et al., A single-pixel terahertz imaging system based on compressed sensing. *Appl. Phys. Lett.*, **93** (2008), 121105.
56. A. Liutkus, D. Martina, S. Popoff et al., Imaging with nature: compressive imaging using a multiply scattering medium. *Sci. Rep.*, **4** (2014). doi: https://doi.org/10.1038/srep05552.
57. J. Hunt and D. R. Smith, Metamaterial apertures for computational imaging. *Science*, **339** (2013), 310–313.
58. E. J. Candes, J. Romberg, and T. Tao, Robust uncertainty principles: exact signal reconstruction from highly incomplete frequency information. *IEEE Trans. Inform. Theory*, **52** (2004), 489–509.
59. D. L. Donoho, For most large underdetermined systems of equations, the minimal l1-norm near-solution approximates the sparsest near-solution. *Comm. Pure Appl. Math.*, **59** (2004), 907–934.
60. D. L. Donoho, Compressed sensing. *IEEE Trans. Inform. Theory*, **52**, 1289–1306 (2006).
61. D. Donoho and J. Tanner, Precise undersampling theorems. *Proc. IEEE*, **98** (2010), 913–924.
62. Y. C. Elad and G. Kutyniok, *Compressed Sensing: Theory and Applications*. Cambridge: Cambridge University Press (2012).

63. M. Elad, *Sparse and Redundant Representations: from Theory to Applications in Signal and Image Processing*. Berlin: Springer (2010).
64. E. Candes and B. Recht, Exact matrix completion via convex optimization. *Found. Comput. Math.*, **6** (2009), 717–772.
65. E. Candes and Y. Plan. Matrix completion with noise. *Proc. IEEE*, **6** (2010), 925–936.
66. J. Mairal, G. Sapiro, and M. Elad, Learning multiscale sparse representations for image and video restoration. *Multiscale Model Simul.*, **7** (2007), 214–241.
67. G. Yu, G. Sapiro, and S. Mallat, Solving inverse problems with piecewise linear estimators: from Gaussian mixture models to structured sparsity. *IEEE Trans. Image Process.*, **21** (2012), 2481–2499.
68. G. Lipworth, A. Mrozack, J. Hunt et al., Metamaterial apertures for coherent computational imaging on the physical layer. *J. Opt. Soc. Am. A*, **30** (2013), 1603.
69. L. Li, F. Li, and T. J. Cui, Feasibility of resonant metalens for the subwavelength imaging using a single sensor in the far field. *Optics Express*, **22** (2014), 18688–18697.
70. L. Li, F. Li, T. J. Cui, and K. Yao, Far-field imaging beyond diffraction limit using single sensor in combination with a resonant aperture. *Optics Express*, **23** (2015), 401–412.
71. L. Wang, L. Li, Y. Li, H. Zhang, and T. J. Cui, Single-shot and single-sensor high/super-resolution microwave imaging based on metasurface. *Sci. Rep.*, **6** (2016), 26959.
72. L. Li et al., A survey on the low-dimensional-model-based electromagnetic imaging. *Found. Trends Signal Process.*, **12** (2018), 107–199.
73. F. Krahmer, S. Mendelson, and H. Rauhut, Suprema of chaos processes and the restricted isometry property. *Commun. Pure Appl. Math.*, **67** (2014), 1877–1904.
74. Y. Xie, T. H. Tsai, A. Konneker et al. Single-sensor multispeaker listening with acoustic metamaterials. *PNAS*, **112** (2015), 10595–10598.
75. F. Lemoult, G. Lerosey, J. de Rosny, and M. Fink, Resonant metalenses for breaking the diffraction barrier. *Phys. Rev. Lett.*, **104** (2010), 203901.
76. F. Lemoult, M. Fink, and G. Lerosey, A polychromatic approach to far-field superlensing at visible wavelengths. *Nat. Commun.*, **3** (2012), 177–180.
77. D. Schurig, J. J. Mock, and D. R. Smith, Electric-field-coupled resonators for negative permittivity metamaterials. *Appl. Phys. Lett.*, **88** (2006), 041109–041109.
78. P. C. Chaumet, A. Sentenac, and A. Rahmani, Coupled dipole method for scatters with large permittivity. *Phys. Rev., E*, **70** (2004), 193–204.

79. P. C. Chaumet and K. Belkebir, Three-dimensional reconstruction from real data using a conjugate gradient-coupled dipole method. *Inverse Probl.*, **25** (2009), 24003–17.
80. M. V. Berry and S. Popescu, Evolution of quantum superoscillations, and optical superresolution without evanescent waves. *J. Phys. A: Math. Gen.*, **39** (2006), 6965–77.
81. E. T. F. Rogers and N. I. Zheludev, Optical super-oscillations: sub-wavelength light focusing and super-resolution imaging. *J. Opt.*, **15** (2013), 094008.
82. E. T. F. Rogers, J. Lindberg, T. Roy et al., A super-oscillatory lens optical microscope for subwavelength imaging. *Nat. Mater.*, **11** (2012), 432–435.
83. A. M. H. Wong and G. V. Eleftheriades, An optical super-microscope for far-field, real-time imaging beyond the diffraction limit. *Sci. Rep.*, **3** (2013), 6330–6337.
84. N. Fang, H. Lee, C. Sun, and X. Zhang, Sub-diffraction-limited optical imaging with a silver superlens. *Science*, **308** (2005), 534.
85. X. Zhang and Z. Liu, Superlenses to overcome the diffraction limit. *Nat. Mater.*, **7** (2008), 435.
86. A. J. Devaney, *Mathematical Foundations of Imaging, Tomography and Wavefield Inversion*. Cambridge: Cambridge University Press (2012).
87. M. Born and E. Wolf, *Principles of Optics: Electromagnetic Theory of Propagation, Interference and Diffraction of Light*. 7th edition. Cambridge: Cambridge University Press (2006).
88. W. Johnson and J. Lindenstrauss, Extensions of Lipschitz mappings into a Hilbert space. *Isr. J. Math.*, **54** (1986), 129–138.
89. I. T. Jolliffe, *Principal Component Analysis*. New York: Springer (2002).
90. M. E. Tipping and C. M. Bishop, Probabilistic principal component analysis. *J. R. Stat. Soc., B: Stat. Methodol.*, **61** (1999), 611–622.
91. N. Halko, P. G. Martinsson, and J. A. Tropp, Finding structure randomness: probabilistic algorithms for constructing approximate matrix decompositions. *SIAM Rev.*, **53** (2011), 217–288.
92. K. Kulkarni and P. Turaga. Reconstruction-free action inference from compressive imagers. *IEEE Trans. Pattern Anal. Mach. Intell.*,**38** (2016), 4, 772–784.
93. S. K. Nayar and V. Branzoi. Programmable imaging: toward a flexible camera. *Int. J. Comput. Vis.*, **20** (2006), 1, 7–22.
94. Hu Tao, A. C. Striwerda, K. Fan et al., Reconfigurable terahertz metamaterials. *Phys. Rev. Lett.*, **103** (2009), 147401.
95. Y.-W. Huang, H. W. H. Lee, R. Sokhoyan et al., Gate-tunable conducting oxide metasurfaces. *Nano Lett.*, **16** (2016), 5319–5325.

96. P. del Hougne, M. F. Imani, M. Fink, D. R. Smith, and G. Lerosey, Precise localization of multiple noncooperative objects in a disordered cavity by wave front shaping. *Phys. Rev. Lett.*, **121** (2018), 063901.
97. K. R. Joshi, D. Bharadia, M. Kotaru, and S. Katti, WiDeo: fine-grained device-free motion tracing using RF backscatter. *Proceeding of the USENIX Conference on Networked Systems Design and Implementation*, (2015), 189–204.
98. X. Dai, Z. Zhou, J. Zhang, and B. Davidson, Ultra-wideband radar-based accurate motion measuring: human body landmark detection and tracing with biomechanical constraints. *IET Radar Sonar Navig.*, **9** (2015), 154–163.
99. Q. Pu, S. Gupta, S. Gollakota, and S. Patel, Whole-home gesture recognition using wireless signals. *Proceedings of the 19th Annual International Conference on Mobile Computing & Networking*, (2013), 27–38.
100. H. Sadreazami, M. Bolic, and S. Rajan, CapsFall: fall detection using ultra-wideband radar and capsule network. *IEEE Access*, **7** (2019), 55336–55343.
101. M. Zhao, T. Li, M. Alsheikh et al., Through-wall human pose estimation using radio signals. *2018 IEEE/CVF Computer Vision and Pattern Recognition*, (2018), 7356–7365.
102. M. Zhao, Y. Tian, H. Zhao et al., RF-based 3D skeletons. *Proceedings of the 2018 Conference of the ACM Special Group on Data Communication*, (2018), 267–281.
103. M. Mercuri, I. R. Lorato, Y. H. Liu et al., Vital-sign monitoring and spatial tracking of multiple people using a contactless radar-based sensor. *Nat. Electron.*, **2** (2019), 252–262.
104. W. P. Huang, C. H. Chang, and T. H. Lee, Real-time and noncontact impulse radio system for μm movement accuracy and vital-sign monitoring applications. *IEEE Sens. J.*, **17** (2018), 2349–2358.
105. S. Ren, K. He, R. Girshick, and J. Sun, Faster R-CNN: towards real-time object detection with region proposal networks. *IEEE Trans. Pattern Anal. Mach. Intell.*, **39** (2016), 1137–1149.
106. D. Huang, R. Nandakumar, and S. Gollakota, Feasibility and limits of Wi-Fi imaging. *Proceedings of the 12th ACM Conference on Embedded Network Sensor Systems*, (2014), 266–279.
107. P. M. Holl and F. Reinhard, Holography of Wi-Fi radiation. *Phys. Rev. Lett.*, **118** (2017), 18390.
108. G. Wang, Y. Zou, Z. Zhou, K. Wu, and M. Ni Lionel, We can hear you with Wi-Fi!, *IEEE Trans. Mob.*, **15** (2016), 2907–2920.

109. Z. Wang, H. Zhang, H. Zhao et al., Intelligent electromagnetic metasurface camera: system design and experimental results, *Nanophotonics*, **11** (2022), 0665.
110. A. Chakrabarti, Learning sensor multiplexing design through backpropagation. *Advances in Neural Information Processing Systems 29: Annual Conference on Neural Information Processing Systems*, (2016), 3081–3089.
111. M. R. Kellman, E. Bostan, N. A. Repina, and L. Waller, Physics-based learned design: optimized coded-illumination for quantitative phase imaging. *IEEE Trans. Comput. Imaging*, **5** (2019), 344–353.
112. V. Sitzmann, S. Diamond, Y. Peng et al., End-to-end optimization of optics and image processing for achromatic extended depth of field and super-resolution imaging. *ACM Trans. Graph.*, **37** (2018), 1–13.
113. J. Chang, V. Sitzmann, X. Dun, W. Heidrich, and G. Wetzstein, Hybrid optical-electronic convolutional neural networks with optimized diffractive optics for image classification. *Sci. Rep.*, **8** (2018), 12324.
114. A. Muthumbi, A. Chaware, K. Kim et al., Learned sensing: jointly optimized microscope hardware for accurate image classification. *Biomed. Opt. Express*, **10** (2019), 6351.
115. S. Vedula, O. Senouf, G. Zurakhov et al., Learning beamforming in ultrasound imaging. *Proc. Mach. Learn.*, **102** (2019), 493–511.
116. D. P. Kingma and J. L. Ba, Adam: a method for stochastic optimization. *arXiv preprint arXiv:1412.6980* (2014).
117. K. He, X. Zhang, S. Ren, and J. Sun, Deep residual learning for image recognition. *IEEE Conference on Computer Vision and Pattern Recognition (CVPR)*, (2016), 770–778.
118. P. del Hougne, M. F. Imani, A. V. Diebold, R. Horstmeyer, and D. R. Smith, Learned integrated sensing pipeline: reconfigurable metasurface transceivers as trainable physical layer in an artificial neural network. *Adv. Sci.*, **7** (2019), 1901913.
119. C. Doersch, Tutorial on variational autoencoders. *arXiv:1606.05908* (2016).
120. L. Li, B. Jafarpour, and M. R. Mohammad-Khaninezhad, A simultaneous perturbation stochastic approximation algorithm for coupled well placement and control optimization under geologic uncertainty. *Comput. Geosci.*, **17** (2013), 167–188.
121. K. Friston, The free-energy principle: a unified brain theory? *Nat. Rev. Neuro.*, **11**, 127–138 (2010).
122. O. Ronneberger, P. Fischer, and T. Brox, U-net: convolutional networks for biomedical image segmentation. *International Conference on Medical*

Image Computing and Computer-Assisted Intervention. (2015), 234–241. www.stereolabs.com/zed-2/.

123. R. Alp Guler, N. Neverova, and I. Kokkinos, DensePose: dense human pose estimation in the wild. *EEE/CVF Conference on Computer Vision and Pattern Recognition*, (2018), 7297–7306. (Code is available at http://densepose.org/.)

Acknowledgments

We would like to express our appreciation to our families for their support, love and encouragement. This Element initially grew out of research projects with our many students. So, we gratefully acknowledge the artwork, discussions, encouragement, feedback, refereeing, and support from Ya Shuang, Hengxin Ruan, Haoyang Li, Zhuo Wang, Libo Wang, Wei Ji, Yunbo Li, and Che Liu. We also thank Cheng-Wei Qiu, Shuang Zhang, and Philipp del Hougne for their support; without it, this Element would not have been written.

Cambridge Elements

Emerging Theories and Technologies in Metamaterials

About the Series

This series systematically covers the theory, characterisation, design and fabrication of metamaterials in areas such as electromagnetics and optics, plasmonics, acoustics and thermal science, nanoelectronics, and nanophotonics, and also showcases the very latest experimental techniques and applications. Presenting cutting-edge research and novel results in a timely, indepth and yet digestible way, it is perfect for graduate students, researchers, and professionals working on metamaterials.

Cambridge Elements ☰

Emerging Theories and Technologies in Metamaterials

Elements in the Series

Spoof Surface Plasmon Metamaterials
Paloma Arroyo Huidobro, Antonio I. Fernández-Domíguez, John B. Pendry, Luis Martín-Moreno, Francisco J. Garcia-Vidal

Metamaterials and Negative Refraction
Rujiang Li, Zuojia Wang, and Hongsheng Chen

Information Metamaterials
Tie Jun Cui and Shuo Liu

Effective Medium Theory of Metamaterials and Metasurfaces
Wei Xiang Jiang, Zhong Lei Mei, and Tie Jun Cui

Epsilon-Near-Zero Metamaterials
Yue Li, Ziheng Zhou, Yijing He, and Hao Li

A Partially Auxetic Metamaterial Inspired by the Maltese Cross
Teik-Cheng Lim

Intelligent Metasurface Sensors
Lianlin Li, Hanting Zhao, and Tie Jun Cui

A full series listing is available at: www.cambridge.org/EMMA

Printed in the United States
by Baker & Taylor Publisher Services